Nanomedicine

A Soft Matter Perspective

Nanomedicine

A Soft Matter Perspective

Edited by
Dipanjan Pan

University of Illinois
Urbana-Champaign

CRC Press
Taylor & Francis Group
Boca Raton London New York

CRC Press is an imprint of the
Taylor & Francis Group, an **informa** business

CRC Press
Taylor & Francis Group
6000 Broken Sound Parkway NW, Suite 300
Boca Raton, FL 33487-2742

First issued in paperback 2019

ISBN-13: 978-1-4665-7282-9 (hbk)
ISBN-13: 978-0-367-37850-9 (pbk)

Library of Congress Cataloging-in-Publication Data

Nanomedicine : a soft matter perspective / editor, Dipanjan Pan.
   p. ; cm.
   Includes bibliographical references and index.
   ISBN 978-1-4665-7282-9 (hardcover : alk. paper)
   I. Pan, Dipanjan, editor of compilation.
   [DNLM: 1. Nanomedicine. 2. Molecular Imaging. QT 36.5]

   R857.N34
   610.28--dc 3   2014004740

Visit the Taylor & Francis Web site at
http://www.taylorandfrancis.com

and the CRC Press Web site at
http://www.crcpress.com

Dedicated to my beloved wife, Angana

Contents

Preface

The capability to monitor disease biosignatures for early and noninvasive detection in combination with targeted therapy is a pressing clinical need and the basis for nanomedicine. This multidisciplinary field is evolving fast and presenting new clinically relevant promises as the science of molecular biology, genomics, chemistry, and nanotechnology contribute greatly. In this book we discuss the unique opportunities presented by biomaterials at the nanoscale. Nanoparticle-based theranostic approaches have emerged as an interdisciplinary area, which shows promise to understand the components, processes, dynamics, and therapies of a disease at a molecular level. The unprecedented potential of nanotechnology for early detection, diagnosis, and personalized treatment of diseases has found application in every biomedical imaging modality. However, with the increasing concern about the ethical and toxicity issues associated with some "nanoplatforms," the biomedical researchers are in pursuit of safer, more precise, and effective ways to practice nanomedicine.

This book provides a broad introduction to matters for nanomedicinal application. It covers comprehensive information regarding "soft" nanoscopic objects with prerequisite features for different imaging modalities. The book also offers a general introduction to the various drug delivery systems and their opportunities from chemistry, materials, biology, and nanomedical standpoints. The book provides a broad introduction to the field and the subfield, with a deeper discussion on the individual modalities for molecular imaging, expanded in four chapters. The theranostic approach is introduced to the readers in the first chapter with discussion about the potential of drug delivery in conjunction with molecular imaging. In Chapter 2, Dr. Ali Azhdarinia and Dr. Sukhen C. Ghosh, from the University of Texas Health Science Center, discuss the potential of nuclear medicine and describe relevant chemical strategies for the design and syntheses of agents suitable for these techniques. In Chapter 3, Dr. Patrick M. Winter, from the Cincinnati Children's Hospital, emphasizes nanomedicine strategies with magnetic resonance imaging. In Chapter 4, Dr. Walter J. Akers, from the Washington University School of Medicine, illustrates various possibilities with optical-based techniques. In Chapter 5, Dr. Dipanjan Pan, Dr. Santosh Misra, and Sumin Kim, from the University of Illinois at Urbana–Champaign, discuss the unique potential for imaging molecular signatures with computed tomography. In Chapter 6, Dr. Dipanjan Pan critically reviews the status of various nanomedicine platforms in clinical trials. This concluding chapter also tries to capture potential pitfalls and challenges faced by the field.

It is anticipated that the book will garner interest across disciplines—within the related areas of chemistry, biochemistry, molecular biology, physiology, and experimental and pharmaceutical sciences. For advanced readers, we expect that it may stimulate experimentation to progress and tune the existing nanomedicine technologies for translational and clinical applications.

The Editor

Prof. Dipanjan Pan joined the University of Illinois in 2013. Previously, he was an assistant professor of medicine, research at the Division of Cardiology, Washington University in St. Louis. He also served as a full faculty member of Siteman Cancer Center at Washington University. After receiving his PhD in chemistry, he pursued a postdoctoral career in polymer science and nanotechnology at the Department of Chemistry, Washington University, in St. Louis. Soon, after a brief stint in the industry (General Electric biosciences/healthcare), Dr. Pan joined the Washington University faculty in 2007.

Prof. Pan's research is broadly aimed at developing clinically translatable defined nanoparticle platforms for molecular imaging, drug delivery, and nonviral gene delivery applications. His research is highly multidisciplinary, which brings skills from synthetic chemistry, nanoengineering, molecular biology, and preclinical animal models. Prof. Pan is a co-inventor of several engineered nanoplatforms for molecular imaging and therapeutic application. His research covers several imaging modalities, including MRI, CT, optical, PET/SPECT, and photoacoustic imaging. His work has been commercialized for preclinical application and externally supported by federal agencies such as NIH, AHA, as well as the Children's Discovery Institute.

Contributors

Walter J. Akers, DVM, PhD
Mallinckrodt Institute of Radiology
Washington University School of
  Medicine
St. Louis, Missouri

Ali Azhdarinia, PhD
Center for Molecular Imaging,
  The Brown Foundation Institute
  of Molecular Medicine
The University of Texas Health Science
  Center at Houston
Houston, Texas

Sukhen C. Ghosh, PhD
Center for Molecular Imaging,
  The Brown Foundation Institute
  of Molecular Medicine
The University of Texas Health Science
  Center at Houston
Houston, Texas

Sumin Kim, BS
Department of Bioengineering and
  Beckman Institute
University of Illinois,
  Urbana–Champaign
Urbana, Illinois

Santosh K. Misra, PhD
Department of Bioengineering and
  Beckman Institute
University of Illinois,
  Urbana–Champaign
Urbana, Illinois

Dipanjan Pan, PhD
Department of Bioengineering and
  Beckman Institute
University of Illinois,
  Urbana–Champaign
Urbana, Illinois

Patrick M. Winter, PhD
Department of Radiology
Cincinnati Children's Hospital
Cincinnati, Ohio

1 Introduction

CONTENTS

1.1 INTRODUCTION

Nanotechnology is the science of manipulating matter on an atomic and molecular scale. Over the past decade, this field has seen enormous growth in terms of precisely manipulating matter at the "nano" scale. At scales on the order of hundreds of nanometers, novel materials properties emerge, enabling the development of new classes of materials. The unique properties experienced at that scale have been utilized to create many new materials with a vast range of applications, such as in electronics, energy production, and, very recently, human health. It can create opportunities for paradigm shifting results, creating preventive, diagnostic, and therapeutic approaches to human diseases such as cancer, neurological, and cardiovascular [1–7] diseases. The strategy for uniting diagnostics and/or therapeutics on the nanoscale is defined popularly as nanomedicine, which offers benefits due to the high degree of equivalent transport and distribution of the active agents within the biological systems for the diagnosis and treatment of diseases.

The biological transport routes, functionally and at the cellular and subcellular levels, are heavily influenced by the physical attributes of the nanoagents. These physical attributes include their size (extra- or intravascular), shape/morphology (e.g., spherical vs. rod), and flexibility ("soft" vs. "hard"), as well as their chemical identities, including surfaces presenting homing agents for specific recognition by and triggering of biological receptors, cell penetrating peptides (CPPs), etc. It is of critical significance to gain understanding of these parameters in details that govern their uniformity with control over their physical and chemical characters.

A subtopic of nanomedicine research is molecular imaging. Emerging trends in molecular imaging (MI) bring promises to recognize the components, progressions, dynamics, and therapies of a disease at a molecular level. MI unites new agents with biomedical imaging tools to identify cellular events visually and depict specific molecular trails in vivo that are key targets in disease progression. The recognition of the existing prospect to detect preclinical pathology has seen myriad progress in this area to synchronous development of sensitive, high-resolution imaging modalities and specific molecular probes.

The most commonly used noninvasive cellular and molecular imaging techniques include clinical modalities—that is, ultrasound (US), positron emission tomography (PET), computed tomography (CT), and magnetic resonance imaging (MRI) and experimental imaging techniques (e.g., optical, photoacoustic imaging [PAI], etc.). Nanoarchitectures for medicinal application can be designed from various precursor materials (e.g., lipids, polymers, metals, etc.). They can act as a reservoir material to encapsulate a wide range of active constituents, including contrast agents, therapeutics, and homing moieties. Soft nanomaterials (e.g., well-defined polymers, lipids, etc.) are excellent examples for their versatility in terms of modifying their structures for high payload and delivery to the disease site. Some of the materials are also known to respond to environmental factors (e.g., physiological or external stimuli). The physiological factors include pH, enzymatic, oxidative, and reductive conditions. The external stimuli can be exemplified as temperature, ultraviolet (UV)-visible (vis) light, near-IR (infrared), stimulation with magnetic fields or ultrasonic vibrations, etc.

The physicochemical properties of these nanoparticles influence their shelf-life and in vivo stability, biodegradability, biocompatibility, biodistribution, and bioelimination. Therefore, the strategy for generating particles for nanomedicine application depends on the target disease, intended site of delivery (organs, tissues, cellular or subcellular organelles), route of administration, and the technique used for imaging. In most cases, intravenous injection is the primary route of administration, but there are other, nonconventional ways of administration (e.g., intradermal/transdermal, oral, and mucosal delivery).

1.2 BIOLOGICAL BARRIERS

For cell- or tissue-specific delivery, drugs are encapsulated into nanoparticles and are modified in a defined way so that they reach their destinations (i.e., sites of action), typically on the cellular or molecular levels. The hurdles of the delivery of these drugs (also their carrier nanoparticles) can be classified into pathway barriers and cellular barriers (the limited cellular uptake, endosomal or lysosomal degradation, and the inefficient translocation to the targeted subcellular organelles). Pathway barriers ("en route" barriers) can be exemplified as the obstacles that nanoparticles experience once they are in blood and extracellular matrix. The behavior of the nanoparticles following intravenous administration is complex. Once in the blood, they are exposed to renal and hepatic clearance, disruption, clumping, opsonization, and clearance by the reticuloendothelial system (RES). RES is now referred to as the mononuclear phagocytic system (MPS). Nanoparticles can extravasate or allocate into the various body tissues and organs, which is highly dependent on their characteristics, size, shape, and composition. The excretion of nanoparticles through the renal pathway is dependent on nanoparticle size. It is believed that a typical renal cutoff is 10 nm or approximately 50 kDa. Hepatic clearance of nanoparticles into bile and feces is another path of excretion from the body. Opsonization is considered one of the major barriers to particle stability in vivo. The examples of opsonins are complement proteins, immunoglobulins, albumin, and fibrinogen. These proteins adsorb on the surface of nanoparticles and label them for attack by the RES. The RES (or MPS) is a part of the human immune system that entails phagocytic cells (e.g., blood monocytes and macrophages accumulated in lymph nodes), spleen, liver (Kupffer cells), and other tissues. These cells scavenge and consume foreign particles when they are identified by the appropriate opsonin. The opsonization of nanoparticles can generate immunological reactions and other in vivo instability to destabilize the particles, leading to the premature release of their payloads.

The mode of cellular uptake is critical for delivering drugs. Nanoparticles may interact with the lipid bilayer of the outer surface of cells and cellular membranes followed by internalization through a process called endocytosis. In this process, the nanoparticles are engulfed into different sizes of vesicles by the plasma membrane. Contingent on the internalization pathway, nanomaterial compositions (lipid vs. polymeric), size, morphology, and surface chemistry, they can create vesicles of different kinds (e.g., endosomes, phagosomes, or macropinosomes). Some of the most commonly documented mechanisms for the uptake of nanoparticles are phagocytosis, lipid-fusion/contact-facilitated, clathrin-mediated, caveolin-mediated, and clathrin/caveolin-independent endocytosis. Translocation to nucleus or mitochondria is another challenging issue due to the physiological nature of these organelles. Recycling (exocytosis) of the vesicle contents is also a possible pathway for the excretion of nanoparticles from cells. Direct translocation of nanoparticles is possible by using a category of peptides called cell-penetrating peptides (e.g., HIV-TAT1). CPPs have been applied for the transport of nanoparticles and cargoes to get a direct access to the cytoplasm after crossing the membrane.

1.3   RATIONAL DESIGN OF NANOPARTICLES

1.3.1   Contrast Materials and Targeting Agents

The term "nano" comes from the ancient Greek νάνος, *nanos*, which refers to "dwarf." It is a unit of measurement and can be denoted as 1×10^{-9} m, or equal to 10 angstroms. At the nanometric scale, due to the size and high surface area, these particles show unique properties. In real-world scenarios, humans are 10^7 times smaller than the Earth, whereas 100 nm sized particles are 10^7 times smaller than humans (Figure 1.1).

These properties are exploited in wide-ranging biomedical applications including imaging and targeted drug delivery. There has been a surge of developing molecular assemblies at the nanometric scale in anticipation that unique chemical and biological properties will be achieved based on their interaction at their increased functional area. The high surface area to volume ratio allows multiple alterations to introduce

- Humans are 10,000,000 times smaller than the Earth.
- A 100 nm sized particle, is 10,000,000 times smaller than a human.

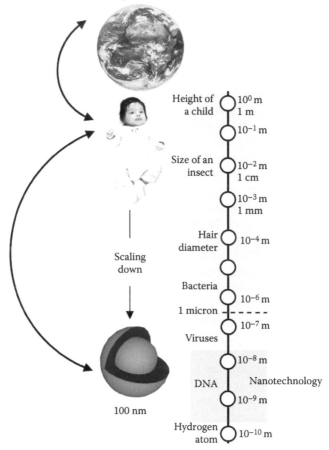

FIGURE 1.1 (**See color insert.**) Size scales in nanotechnology.

homing agents, contrast agents, and drug payloads internally or on the surface. Thus, the number of ligands per nanoparticle is increased, which in turn improves their binding affinity and decreases the rate of dissociation. This "stick-and-stay" effect is frequently referred to as avidity of a nanoparticle.

Toward this goal, well-defined nanoscale materials are being developed involving conventional and advanced chemical strategies to carry contrast materials (e.g., radio isotopes ^{99m}Tc, paramagnetic/superparamagnetic metals, fluorophores, heavy metals, etc.). Other constructs are, themselves, the image-able agent, but are specifically tuned for targeting (e.g., dextran-coated iron oxides).

For developing clinically translatable nanomedicine agents, the materials must possess a tolerable toxicity (acute, systemic, immune response, cellular), long circulating half-life, excellent contrast-to-noise ratio enhancement, high binding efficacy, and compatibility with commonly used commercial imaging modalities.

The electronic structure and reactivity of a material changes as the particle size changes and, at such very small scales, nanoparticles are dominated by their surfaces. Research at nanometric scale has shown that size of the particles really matters and governs their accessibility to targets and their rate of clearance from tissues. The definition of nanoparticle in the biological sciences is dynamic, referring to particles generally less than 500 nm. According to the definition of the National Cancer Institute, a nanoparticle can be defined as "a particle that is smaller than 100 nanometers (one-billionth of a meter). In medicine, nanoparticles can be used to carry antibodies, drugs, imaging agents, or other substances to certain parts of the body. Nanoparticles are being studied in the detection, diagnosis, and treatment of cancer" (NCI Dictionary of Cancer Terms: http://www.cancer.gov/dictionary?cdrid=653131). Indeed, agents larger than 100 nm are more appropriate for intravascular targets, whereas their smaller counterparts follow endothelial permeability and retention mechanisms to be directed to the diseased tissues. The next section will highlight the choice of various nanoparticle platforms and the synthetic strategies involved.

1.3.2 TARGETING STRATEGIES AND HOMING AGENTS

In general, passive targeting can provide microanatomical information. In order to obtain specific biochemical information, active targeting strategies are required to provide molecular specificities. Passive targeting relies on the characteristics of the delivery vehicle and the pathology of the disease for the preferential accumulation of nanoparticles at the site of interest. Nanoparticle uptake by passive mechanism is commonly recognized as enhanced permeability and retention (EPR) effect due to the hyperporousness of the leaky blood vessels. It was first reported by Maeda and colleagues [8–10]. Other approaches of passive targeting involve using stimuli-responsive materials for accumulating therapeutics to the diseased sites [11–15]. Some nanoparticles can be directly injected into the tumors for therapeutic delivery and detection. As a specific example, gold nanoparticles can be directly injected into the tumor site and then heated via IR for thermal therapeutics effects [16].

Active targeting strategies to the disease tissues involve chemical conjugation of a specific moiety on the surface of the nanomaterials that will be identified by the specific cells present at the disease site. A wide variety of targeting moieties are utilized to decorate ligand-directed nanomedicine agents (e.g., monoclonal antibodies [and F_{ab} fragments] [17–19], small molecule ligands [20–23], peptidomimetics [24,25], aptamers [26–28], and polysaccharides [e.g., heparin] [29–32]). Various chemical strategies can be adopted to attach these ligands to the nanoparticles as described in Section 1.3.4.

In order to prolong the circulation time of these materials in vivo, some agents often carry a polyethylene glycol (PEG) functionality attached to the surface of these particles. PEG or a short carbon-chain caproyl type spacer can also be introduced within the homing moiety and the nanoparticles to assimilate the surface in three-dimensional spaces freely. This modification also facilitates the actual cellular binding event to ensue by positioning the moiety away from the surface of the particle in an orientation preferable for recognition by the receptor.

1.3.3   Nanoparticle Platforms

A plethora of chemical approaches has been proposed for nanocarrier platforms designed for proof-of-concept preclinical imaging and for active and passive delivery of therapeutics. A generalized illustration of a multifunctional nanoparticle platform is shown in Figure 1.2. The type of the precursor material used for these platforms is dominated by "hard" materials (crystalline metallic). As a specific example, MRI probes are typically based on iron oxides, in magnetite and maghemite forms. Superparamagnetic iron oxides (SPIOs) constitute iron oxide, coated with hydrophilic agents (e.g., polymers [PMMA, PVA, PLGA, PEG, PEG-lipid], dextrans, starch, oleic acid, etc.). SPIO can be categorized as standard SPIO (SSPIO) with particle sizes ranging from 300 nm to micron level and relatively small SPIO (50–150 nm). Ultrasmall SPIO (USPIO) and monocrystalline iron oxide nanoparticles (MION) represent the tiniest of the iron oxide family with sizes ranging from 10 to 40 nm. These particles are superparamagnetic in nature and have shown the earliest known applications of MR imaging in healthy hepatocytes [33], prostate cancer [34], and even in atherosclerotic rabbit models [35].

The examples of "soft" nanocarriers include lipid-based and polymeric materials. Examples of dendrimers [36,37], shell-cross-linked polymers [38], nanocages [39–41], magnetomicelles [42], virus particles [43,44] and ultrasmall superparamagnetic iron oxide (USPIO) [45–50] are well known. Dendrimers have been studied extensively in vitro and in vivo [36,37]. Star polymers [51–53] and other related structures [54,55] have demonstrated that dynamic high-resolution micro-MRI contrast agent ("dendrimer-enhanced MRI") can be a powerful tool for in vivo observation of renal structure, function, and injury. These agents can also serve as novel markers for diagnosis and prognosis of sepsis-induced acute renal failure in aged mice.

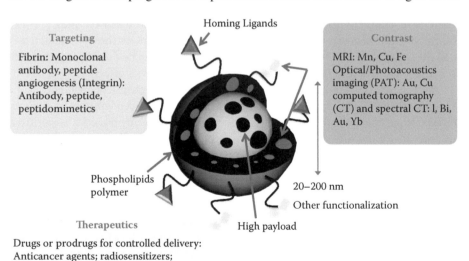

FIGURE 1.2 **(See color insert.)** A cartoon of multifunctional nanoparticles representing an extremely versatile platform for molecular imaging and drug delivery application.

For targeted imaging, gadolinium (Gd)-labeled, avidin-conjugated dendrimer probes have been efficiently internalized into SHIN3 ovarian adenocarcinoma cells in vitro and shown to be accumulated in the intraperitoneally disseminated ovarian tumor (SHIN3) mouse model [56]. Dendritic probes have been successfully employed in vivo for assessing articular cartilage by MR [57] and in angiogenesis [58,59].

A lipid-based particle system is another attractive choice as nanomedicine platform due to its biological inertness, cellular fusion, and easy access to synthetic routes. Liposomes [60–66] and phospholipid-encapsulated nanoemulsions [67,68] have been studied comprehensively for targeted imaging and therapeutic applications. One such example is the perfluorocarbon (PFC) nanoparticle, which has been applied to clinically relevant modalities [68(a)] for imaging and its potential for drug delivery capabilities has been well established. Ligand-directed, phospholipid-coated PFC nanoparticles are vascularly constrained (150–240 nm) and fluidic (flexible) in characteristics. The surface of these nanoparticles may carry at least 100,000 contrast metals and several hundreds of targeting moieties per particle, which provides excellent signal amplification in vivo and increased avidity (i.e., "stick and stay") quality. Consequently, the surface area to volume ratio permits an even better concentrating effect to transport substantial amounts of probes per binding site and per volume, which allows them to be detected sensitively by imaging techniques even for sparser biological targets (e.g., angiogencsis). The capabilities of these particles can further be stretched to embrace drug delivery. The multifunctional capabilities of these agents would simultaneously allow confirmation and quantification of local delivery for the response to be serially monitored for targeted agent. The utility of these agents for therapeutic delivery, which can also be imaged for diagnostic or dose quantification purposes, has been successfully demonstrated [68(b)–(d)]. Due to the dual capabilities/function of these particles, they are often referred to as "theranostic" (therapy and imaging) or "theragonostic" agents.

1.3.4 SURFACE CHEMISTRIES

Targeting moieties are attached to the particles by different chemical reactions, which may involve noncovalent interactions or covalent conjugation through reactive functionalities. As a specific example of noncovalent conjugation strategy, biotin–avidin interaction has been used to couple homing agents to nanoparticles in an interchangeable fashion. Avidin and streptavidin are protein molecules containing four binding holes for biotins. They have been successfully used in vitro and in vivo as noncovalent ways for targeting vascular epitopes [69,70].

Bioconjugation techniques are getting increasingly prevalent for fast and consistent covalent coupling of ligands [71]. Carbodiimide-coupling chemistry is one of the most widely used techniques for conjugating antibodies or amine terminated ligands with particles presenting carboxyl groups on the surface [21,25]. The other most commonly used techniques frequently involve active ester coupling, acylation, alkylation, and click-chemistry protocol [72]. Likewise, nanoparticles presenting amines are coupled to the carboxyl [73] or aldehyde [74] terminated ligands to form amide or imine linkages. Peptides, oligonucleotides, or other small molecules are typically attached to the sulfohydryl-modified nanoparticles through disulfide bond formation [75]. Nanoparticles presenting hydroxyl moieties can be activated to form a reactive species, which freely reacts with amine terminated ligands and antibodies forming a "zero-length" spacer [76].

Carbodiimide-mediated coupling chemistries are possibly the most widely explored bioconjugation protocol. The reaction is straightforward; however, several possible side reactions confound the subject. The side reaction of the intermediate O-acylisourea produces both desired and undesired products. O-acylisourea suffers undesired rearrangements to the stable and nonreactive intermediate N-acylurea. Bioconjugation techniques are water mediated and this helps to avoid the use of non-polar solvents with low-dielectric constants (e.g., dichloromethane or chloroform) minimizing the side reaction. Usually, EDC (also EDAC or EDCI, 1-ethyl-3-(3-dimethylaminopropyl) carbodiimide hydrochloride), a highly water-soluble car-bodiimide is employed in the 4.0–6.0 pH range to activate carboxyl groups for the coupling of primary amines to produce amide bonds. To enhance the coupling efficiencies, it is often used in combination with N-hydroxysuccinimide (NHS) or sulfo-NHS. EDC can also be used to activate phosphate groups and finds widespread application in formulation of immunoconjugates, solution and solid phase-peptide synthesis, protein cross linking to nucleic acids, etc.

Synthetic methodologies based on "click" reactions [77] are gaining popularity quickly. "Click-chemistry" techniques have wide acceptance mainly due to high reliability, tolerance to a broad variety of functional groups, quantitative yields, and their usefulness under mild conditions. The capability to reduce Cu catalyst in situ has now permitted "click chemistry" to be achieved in aqueous media, which resolves the oxidative instability of the catalyst under aerobic environments. Click-chemistry protocols have been employed for bacterial cell surface labeling [78] and in conjugation of biological polymers to viruses [79,80], synthetic polymers, and others [81–85].

1.4 SPECIFIC EXAMPLES OF TARGETED MOLECULAR IMAGING AND THERAPY

Research is being conducted in the field of molecular imaging with sensitive, high-resolution imaging modalities. Molecular imaging describes tools that allow visualization and quantification over time of signals from molecular imaging agents and employs clinically applicable imaging techniques such as US, PET, CT and MRI. Modalities provide unique strengths and their applications are tailored based on the clinical need. For example, for coronary scanning, computed tomography is preferred, since the image acquisition is much faster than MRI. The combination of these different modalities should allow better diagnostics and therapies such as early stage disease diagnosis, drug delivery, image-guided surgery, and cancer staging. A nanomedicinal approach can also be applied to resolve intractable problems of drug instability and premature release of therapeutics in transit to target sites. For example, antiangiogenic fumagillin suffers from its poor retention during intravas-cular transit and requires an innovative solution for clinical translation. An Sn-2 lipase-labile fumagillin prodrug was developed and delivered as an antiangiogenic

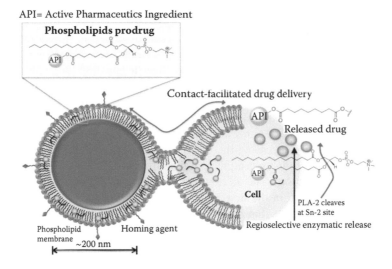

FIGURE 1.3 **(See color insert.)** A cartoon depicting the mechanistic concept of a therapeutic delivery via "contact facilitated drug delivery" mechanism.

prodrug, in combination with a contact-facilitated drug delivery mechanism [11(b)] (Figure 1.3).

Studies are being conducted to improve the imaging techniques in terms of better resolution, in vivo depth penetration, and safety (e.g., less radiation). In this chapter we will particularly emphasize advanced imaging methods and targeted probes for US and MRI and discuss some evolving imaging techniques (Figure 1.4).

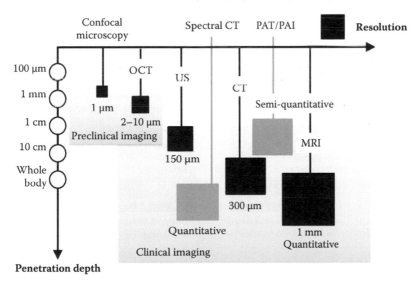

FIGURE 1.4 Commonly used clinical and preclinical imaging modalities in an overlapping scale depicting their comparative image resolution and depth penetration.

1.4.1 OPPORTUNITIES WITH ULTRASOUND IMAGING

Ultrasound is the most extensively used clinical imaging modality [86,87]. Ultrasonography has long been employed for applications such as blood pool enhancement, perfusion imaging, and classification of liver lesions [88–90]. Commercially available, gas-filled encapsulated microbubbles (~2–5 μm) have been known for years as blood pool contrast agents [91]. The use of targeted microbubbles for integrin $\alpha_v\beta_3$ and vascular endothelial growth factor receptor 2 (VEGFR-2; Flk-1/KDR) has been described [92–97]. However, in 2007, the Food and Drug Administration (FDA) issued an alert concerning the use of microbubble-based injectable suspension (e.g., Perflutren lipid microsphere) and Optison (Perflutren protein-type A microspheres for injection). The alert stressed the risk of serious cardiopulmonary responses during or within the first 30 minutes following the administration. The recommendation was made that high-risk patients with pulmonary hypertension or unstable cardiopulmonary conditions be closely monitored. However, in 2008, some of these cautions (*contraindications*) were removed because the FDA determined that, in some patients, the benefits from the diagnostic information that could be gained through the use of these agents may offset the danger of serious cardiopulmonary reactions, even among some patients at particularly high risk for these reactions. One of the other major disadvantages of using microbubbles for targeted imaging is their high contextual signal and "tethering" of these particles to a surface, which limits their capability to fluctuate, and thus their echogenicity is hindered (Figure 1.5).

A novel multifunctional lipid encapsulated nongaseous PFC emulsion was reported addressing these issues with microbubbles [68]. This material was vascularly constrained by size (150–250 nm) and robustly stable to usage, pressure, heat, and shear forces. PFC nanoparticles are smaller than microbubbles and more stable in vivo. They have a long circulatory half-life and they are untraceable in circulation. Furthermore, PFC nanoparticles are not PEGylated on their outer surface, which is critical to preserve their targeting efficacies. These materials have lowly inherent acoustic reflectivity and have been found to possess backscattering levels below the level of whole blood [98]. However, a collective buildup of these particles on the surface of biological tissues creates a local acoustic impedance mismatch that generates

Control emulsion   Biotinylated emulsion
(a)   (b)

FIGURE 1.5 Ultrasonic images (7.5 MHz) of plasma clots pretargeted with antifibrin monoclonal antibody and exposed to control (left) or biotinylated PFC nanoparticle (right) in vitro. The supporting suture (s) and thrombus (t) are clearly delineated. As evident from the images, (a) the control emulsion did not enhance the acoustic reflectivity of the clots, whereas (b) the biotinylated PFC nanoparticles greatly increased the echogenicity of the targeted clot (b).

a strong ultrasound signal (Figure 1.4). Due to their nongaseous nature, they are not easily deformed or cavitate during ultrasound imaging; thus, an ultrasound signal can be obtained for long periods [99]. For targeted imaging, ligand-coated PFC emulsion nanoparticles have been developed and used to identify the angioplasty-induced tissue factor by smooth muscle cells within the tunica media [100,101]. Similar PFC nanoparticles targeted to markers of angiogenesis have been used successfully to detect neovasculature around tumors implanted in athymic nude mice using a research ultrasound scanner [102]. Unlike a nanoemulsion-based platform, a polymeric microparticle derived from poly(lactide-coglycolide) showed in vivo contrast-enhanced power Doppler ultrasonography evaluation on melanoma grafted mice with 50% enhancement of detection of intratumoral vascularization after injection of 3 mm [103].

1.4.2 Opportunities with MR Imaging

Magnetic resonance imaging is a noninvasive diagnostic technique based on the interaction of nuclei with each other and with surrounding molecules in a tissue within externally applied magnetic fields of interest. Although the sensitivity of MR is low in comparison to nuclear and optical modalities, the absence of radiation (transmitted or injected) and high spatial resolution (e.g., submillimeter) make it advantageous over the techniques involving radioisotopes. The inherent low sensitivity is also now being somewhat compensated with the advent of higher magnetic fields (4.7–14 T). Although there are many medically relevant nuclei, the ^1H nucleus (i.e., proton) has been the most widely studied in clinical practice due to its high gyromagnetic ratio and high abundance in nature and biological tissues. ^{19}F has a similar high gyromagnetic ratio as proton and a very high (~100%) natural occurrence that, make it an interesting choice for MR imaging [104]. At constant field strength and with equivalent numbers of nuclei, ^{19}F has sensitivity of about 83% compared to ^1H. In MRI, the protons in different tissues have different relaxation times, which can result in endogenous image contrast. Indeed, the initial driving application for clinical MRI was the demonstration of different relaxation times in cancerous tissue versus normal tissue [105]. The inherent image contrast can be enhanced by using exogenous contrast agents preferentially to shorten either the T1 (longitudinal, spin–lattice) or T2 (transverse, spin–spin) relaxation time. Relaxivity can be defined as the ability of contrast agents to increase the relaxation rates of the surrounding water proton spins. The MR image can be weighted to detect differences in either T1 or T2 by adjusting parameters during data acquisition. Gadolinium and iron oxide-based agents have dominated the MR contrast agent scenario and have been used to increase the T1- and T2-weighted image contrast [106–109]. T1 agents influence protons within their proximity and are highly dependent on local water flux [110] predominantly to shorten T1 relaxation to produce bright signals on T1-weighted images. Iron oxide-based agents, on the other hand, can be considered as T2 agents, which exhibit high magnetic susceptibility and produce local disturbances in the field [111]. Signal dephasing in tissues results from these magnetic susceptibility heterogeneity disturbances, which leads to a loss of signal through T2* decay, especially on gradient echo images. Nontargeted iron oxide-based agents are normally taken up nonspecifically by the RES and their biodistribution is governed by the overall size of the nanoparticle [112–118]. The examples of passive targeting with iron oxide-based agents include brain tumors [119], stroke [120], angiogenic tumor vessels [121–123], multiple sclerosis [124], carotid atherosclerotic plaque [125,126], and more. Iron oxide nanoparticles have recently been used for trafficking cells in vivo with MRI [127–131] owing to their low detectability limit [132].

Manganese was one of the first reported examples [133–135] of paramagnetic contrast material studied in cardiac and hepatic MRI because of its efficient R1 (1/T1) enhancement. Similarly to Ca^{2+} and unlike the lanthanides, manganese is a natural cellular constituent and often a cofactor for enzymes and receptors. Manganese blood pool agents, such as mangafodipir trisodium, have been approved as hepatocyte-specific contrast agents with transient side effects due to dechelation of manganese from the linear chelate. Manganese-enhanced magnetic resonance imaging (MEMRI) has long been known to provide unique information in modern neuro-imaging. Combination of the rich biology of Mn^{2+} and its properties as an MRI contrast agent allows MEMRI to be uniquely useful for the detection of anatomical structures after its systemic administration. Also, MEMRI is known for mapping functional brain regions by monitoring the dynamics of Mn^{2+} entry into excitable cells in the central nervous system (CNS), which is intrinsically linked to activity of voltage-gated calcium channels [136–139]. Nontargeted liposomal agents have included $MnSO_4$ [134] or Mn-DTPA [135]. Release of manganese caused by disruption of the vesicles allowed MR detection of sites where the vesicles were nonspecifically entrapped. A major drawback to the use of manganese as a contrast agent, however, is its cellular toxicity. Therefore, its successful implementation as a contrast agent depends on the site-specific delivery at a dose as low as possible, maintaining the detectability by MRI.

The challenge to overcome the sensitivity and partial volume effect that is normally being faced by most of the T1- and T2-weighted MR contrast agents prompted the development of T1-weighted paramagnetic PFC particles. Typically, for a 200 nm diameter particle, more than 100,000 gadolinium atoms per particle can be incorporated within the outer surface of the lipid membrane, where they influence the surrounding water for maximum T1 relaxation effect [140,141]. In doing so, a high payload of gadolinium is incorporated to produce an adequate MR signal. The relaxivity of reported [142] Gd^{3+}-chelates (4.5 mM^{-1}sec^{-1}) is lower than Gd^{3+} bound to the surface of the particle (33.7 mM^{-1}sec^{-1}) as measured at 1.5 T [141]. The particulate relaxivity (i.e., the relaxivity on a per-particle basis, rather than per concentration of Gd) has also been found to be very high (~2,000,000 mM^{-1}sec^{-1}) since particles carry a heavy payload of paramagnetic metal atoms, making them better agents for targeted molecular imaging and not as blood pool contrast agents. This has been exploited to image quantitatively and successfully predict a range of sparse concentrations in experimental phantoms and targeted smooth muscle cell (SMC) monolayers expressing "tissue factor" [143]. Unlike paramagnetic gadolinium-based particles carrying a high payload of metals, gadofosveset trisodium (MS-325) is a small molecule contrast agent that has a strong affinity toward albumin, and it is highly protein bound after injection providing vascular signal enhancement superior to that of other agents [144]. Notably, MS-325 is currently being tested under clinical trials and could possibly be the first gadolinium-based blood pool MR imaging agent to progress to human trials [144].

The recent discovery of nephrogenic systemic fibrosis (NSF), a serious and unexpected side effect of gadolinium blood pool agents observed in some patients with renal disease or following liver transplant, has cast a shadow on currently approved blood pool contrast agents and created a new regulatory barrier for targeted agents to address [145–147]. The cause of NSF is unknown and there is no effective treatment of this condition. Given the association between gadolinium and nephrogenic systemic fibrosis, there has also been a growing interest to explore other metals for MR imaging.

Hyeon and co-workers have recently reported a manganese oxide-based targeted T1 MRI contrast agent for T1-weighted MR imaging of various body organs [148]. Pan et al. have reported a new class of manganese (III)-labeled, toroidal-shaped, vascularly constrained (~180 nm) nanoparticles, "nanobialys (MnNBs)," for manganese-based MR imaging [149]. An inversion recovery sequence was used to calculate the ionic (per metal) and particulate (per particle) relaxivities of serially diluted nanobialys at 1.5 T and 25°C. The high particulate relaxivities of the MnNB, $r1 = 612,307 \pm 7,213$ and $r2 = 866,989 \pm 10,704$ (s·mmol [nanobialy])$^{-1}$ measured at 1.5 T (25°C), with ionic $r1$ and $r2$ relaxivities of 3.7 ± 1.1 and 5.2 ± 1.1 (s·mmol [Mn])$^{-1}$, points to a novel fibrin-specific nongadolinium, molecular imaging platform that offers a sensitive noninvasive MR imaging approach for diagnosis of rupturing atherosclerotic.

The development of activatable MR contrast agents was first reported by Thomas Meade and was developed to correlate developmental biological events with gene expression during an imaging experiment [150(a),(b)]. His paper described the in vivo MR imaging of the β-galactosidase marker gene and demonstrated the potential of MRI to track gene expression. The "blocking" galactopyranose sugar residue is cleaved by β-galactosidase and the relaxivity of the agent is enhanced, allowing water to access the inner coordination sphere of the agent. As a new class of environmentally responsive MR contrast agents, very recently, Shapiro and Koretsky described the use of inorganic manganese-oxide and carbonate-based particles as convertible MRI agents. These particles exhibited only susceptibility-induced MRI contrast, most often seen as dark contrast in susceptibility-weighted images. The selective degradation of these particles within the endosomal and lysosomal compartments of cells prompted the dissolved Mn^{2+} to act as a T1 agent. The concept of environmentally responsible MR contrast is demonstrated both in vitro and in vivo, in rat brain [150(c)].

The acute formation of thrombus following atherosclerotic plaque rupture has been well recognized as the etiology of unstable angina, myocardial infarction, transient ischemic attacks, and stroke following the seminal works of Benson [151] and Constantinides [152]. The most common source of thromboembolism remains a rupturing thin cap fibroatheroma (TCFA), which is characterized by a necrotic core with an overlying fibrous cap measuring <65 μm [153–156]. In patients dying of acute myocardial infarction, ruptured TCFA is frequently observed in the proximal coronary vessels with only 50% to 60% residual stenosis [157,158]. In at least 50% of patients with acute ST elevation myocardial infarction, coronary thrombi were days or weeks old, indicating that sudden coronary occlusion often follows a variable period of plaque instability and thrombus formation [159,160]. Furthermore, subclinical episodes of plaque disruption followed by healing contribute to the phasic rather than linear progression of coronary disease observed in angiograms carried out serially in patients with chronic ischemic heart disease [161,162]. Sensitive detection, quantification, and differentiation of intravascular thrombus in vessels with mild severity stenosis could provide a direct metric to risk stratify patients to aggressive interventional versus conservative medical management care strategies. While invasive imaging techniques such as intravascular ultrasound [163], optical coherence tomography [164], and direct angioscopy [165] can identify these lesions, from a practical perspective, a noninvasive imaging approach is required.

Magnetic resonance imaging has emerged as a particularly sensitive, nonionizing modality to visualize thromboses within the carotid artery noninvasively [162]. High-resolution MRI detection and characterization of atherosclerotic lesions (including advanced lesions, such as the fibrous cap, the lipid core, and even plaque fissuring) with serial imaging over time has been used to assess lesion progression or regression [166–169]. High-resolution MRI can be used to distinguish intact, thick fibrous caps from intact thin and disrupted caps in atherosclerotic human carotid arteries in vivo [170,171]. Advances in high spatial resolution black-blood techniques further improved noninvasive imaging of human coronary and carotid arteries and facilitated the early assessment of atherosclerotic disease [172]. Although cardiorespiratory motion has complicated coronary imaging with MRI; rapid advancements in motion correction are being reported, which will likely limit or resolve this issue in the not-too-distant future [173–175]. Sensitive and fibrin-specific contrast agents based on PFC nanoemulsions [140,176] and peptides [177] have recently been developed that are capable of detecting and quantifying microthrombus.

As briefly discussed before, perfluorocarbon nanoparticles have been coupled with paramagnetic atoms to create a bifunctional contrast agent that can be used as a paramagnetic MRI agent separately and in combination with fluorine imaging. In one of the earliest studies [140] with targeted MRI agents, a lipid-conjugated gadolinium–DTPA complex was incorporated into the surfactant layer and demonstrated the result in vitro, and targeting artificially prepared plasma clots in dogs in situ. Subsequently, the MRI agent was refined to optimize the concentration of gadolinium borne on the particle surface [178–181]. Thus, an agent was developed that can overcome the partial volume effects that have negated so many previous attempts at targeted MRI agents. In this experiment using MRI, the 2-0-20 gadolinium emulsions (2% lipid component and 20% of the PFC core) provided outstanding contrast properties and overcame the partial volume effects expected with routine imaging at 1.5 T when targeted to artificially prepared fibrin clots. Although the influence of the gadolinium agent appears to affect the entire clot, high-resolution images reveal that these marked reductions in T1 values are apparent, despite binding of only a single thin layer of the contrast agent.

Morawski et al. have demonstrated the utility of fibrin-targeted nanoparticles to delineate intravascular thrombus in dogs and conceptually on vulnerable plaque using human carotid endarterectomy (CEA) specimens from patients with symptomatic stenosis, transient ischemic attack (TIA), or stroke [179,180]. These data indicate that small, perhaps even microscopic, deposits of human fibrin in micro-fissures of vulnerable carotid plaques can be detected with the contrast agent. The results further suggest that thrombus produced by local plaque rupture or embolic derived could be targeted and imaged, allowing the localization of arterial occlusion and its resolution to be monitored. PFC nanoparticles were further exploited for ^{19}F MR imaging and spectroscopy owing to their high fluorine content. Morawski et al. [143] uniquely quantified the number of nanoparticles bound to the fibrin clot using the ^{19}F signal [182–183].

This leads us to the feasibility for ^{19}F imaging of both perfluorooctylbromide (PFOB) and perfluoro-15-crown-5 ether (CE) nanoparticle-labeled stem/progenitor cells. At present, very few noninvasive techniques exist for monitoring cells after administration and therefore cellular therapeutics show great promise for the treat-ment of disease. Historically, histological techniques have been employed for mon-itoring of therapeutic cells and require sacrifice of the animal or tissue biopsies. Several MRI-based agents (e.g., gadolinium [184,185], superparamagnetic iron oxides [186–193], and a few others [194]) find significant use to visualize immune cells and others in vivo. However, contrast agents based on protons suffer from high background signals from water, which makes it difficult to identify the transplanted cells unequivocally in vivo. Fluorine MRI imaging of cells labeled with ^{19}F-labeled contrast agents may present a unique and sensitive technique that is quantifiable and distinct from any tissue background signal. Ahrens et al. reported efficient intra-cellular uptake of the perfluoropolyether (PFPE)-based agents by phenotypically defined dendritic cells (DCs) ex vivo into tissue or intravenously in mice and then tracked the cells in vivo using ^{19}F MRI [195]. Along the same line, Partlow et al. recently demonstrated that PFC nanoparticles provide an unequivocal and unique signature for stem/progenitor cells, enabling spatial cell localization with ^{19}F MRI, and they permit quantification and detection of multiple fluorine signatures with fluorine MR imaging and spectroscopy [196]. PFC-labeled cells were detectable rap-idly, not only in vitro, but also in liver tissue in vivo with 11.7 T spectroscopy and at potential target sites in situ using both experimental (11.7 T) and clinical (1.5 T) field strengths. The nanoparticle comprised the 20% (v/v) PFC with either PFOB or CE core. Study revealed that both PFOB and CE were readily internalized by stem/progenitor cells without the need for additional use of transfecting agents. Similar methods previously performed using iron oxide particles can potentially lead to sig-nificant losses in cell viability [197]. The resultant intracellular fluorine levels were readily detectable at both 11.7 and 1.5 T, considering the relatively low number of cells utilized and the reasonably short imaging times (under 7 min) [196]. The ability to quantify fluorine levels present and to differentiate distinct PFC signals with MR spectroscopy could provide additional advantages such as the type and number of cells accumulating at target sites.

PARACEST (paramagnetic chemical exchange saturation transfer) agents rep-resent a new class of MRI contrast agents that has been recently introduced, where the bulk water signal intensity is altered via chemical exchange saturation trans-fer (CEST), which was later extended to paramagnetic systems [198]. PARACEST agents are ideally suited for molecular imaging applications because one can switch the contrast on and off as desired simply by adjusting the pulse sequence parameters. This avoids the extra need to record pre- and postinjection images to define contrast agent binding. Liposome-based chemical exchange saturation transfer (lipoCEST) agents have shown great sensitivity and potential for molecular MRI [199–203].

A lipid-encapsulated PFC nanoparticle-based molecular imaging agent that utilizes a PARACEST chelate was developed and reported recently [204,205]. Nanoparticles produced >10% signal enhancement and a bound water peak at 52 ppm, in spectroscopy (4.7 T) of PARACEST nanoparticles, which is in agreement with results from the water-soluble chelates. PARACEST nanoparticles were targeted to artificially prepared plasma clots via antifibrin antibodies, which produced a contrast-to-noise ratio (CNR) of 10 at the clot surface [205].

1.4.3 Opportunities with Emerging Imaging Technologies

Photoacoustic (PA) imaging, also known as optoacoustic imaging, is an evolving hybrid optical and ultrasound diagnostic modality with high spatial resolution and outstanding soft tissue contrast [206–212]. In this technique biological tissue is irradiated with a non-ionizing short-pulsed laser beam within acceptable limits recognized by the American National Standards Institute (ANSI). Endogenous proteins (e.g., hemoglobin, myoglobin, melanin) absorb the optical energy to undergo thermoelastic expansion and radiate acoustic waves that are detected with a clinical, wide-band ultrasonic transducer (5–50 MHz).

Research in this area over the years has shown that PAI has the ability to provide both physiological and molecular imaging data, which can be used alone or in a hybrid modality fashion to extend the anatomic and hemodynamic sensitivities of clinical ultrasound. PAI may be performed using contrast afforded by light absorbing endogenous molecules or exogenous contrast agents. The examples of exogenous small molecule contrast agents include near infrared dyes, porphyrins, and nanoparticles. Myriad advances have been made in terms of exogenous contrast agent development for PA molecular imaging applications, particularly nanoparticle-based technologies. Several optically active materials have been explored as PA contrast agents (e.g., small molecule dyes, gold nanoparticles, single-walled carbon nanotubes [SWNTs], and copper nanoparticles) [213–216]. Beyond needing to exceed the inherent contrast of blood and muscle, these agents must have outstanding biocompatibility, appropriate in vivo stability (i.e., particularly during circulatory transit to targets and through the imaging period), and desirable synthetic processing attributes—specifically, tolerance to sterilization and prolonged shelf-life stability.

Gold nanoparticles, which are excitable in the near-infrared (NIR) range within the so-called "optical transmission window" of biological tissues (λ_{max} = 650–900 nm) afford strong optical absorption, unlike small molecule fluorophores, and are resistant to photobleaching or fading due to dye destruction [217–219]. Gold particles exhibit localized surface plasmon resonance (LSPR) due to their inherent abilities to absorb and scatter light at specific wavelengths. Nanoparticle LSPR is established when incident light photons match the frequency of particle surface electrons vibrating against the opposing restoring force of positive nuclei. Importantly, LSPR can be favorably manipulated by altering particle shape and surface coating.

Gold nanobeacons (GNBs) entrap numerous small gold nanoparticles (spherical, 2–4 nm) within a larger phospholipids-encapsulated colloidal particle [206–212,220]. The entrapment of numerous small gold nanoparticles amplifies the PA signal for each binding event by effectively mimicking a larger single gold particle. However, from a safety perspective, the metabolism of GNB releases tiny particles within the direct renal elimination window (6–10 nm), whereas large gold particles would be retained indefinitely. Whether such long-term retention is "unsafe" or "benign" is unfortunately not predictable and not provable to a risk-averse regulatory agency. Further, the random interactions of the small particles within the particle matrix core effectively create irregular shapes that shift the optical absorbance peak from the visible into the NIR spectral region for improved in vivo use.

Although photoacoustic tomography generates high-resolution images of red blood cells in the microvasculature [206–210], hemoglobin imaging cannot discriminate immature neovasculature from mature microvessels. Moreover, neovessel sprouting and bridging, which is a significant component of the neovascular contrast signal, typically lacks significant blood flow and is otherwise invisible to PA imaging. Fortunately, the steric constraint of GNB and other particles exceeding 100 nm diameter within the vasculature precludes significant interaction with nonendothelial integrin-expressing cells and greatly enhances neovascular target specificity. $\alpha_v\beta_3$-targeted GNB_1 (160 nm), as opposed to GNB_S (90 nm) with a higher potential for extravasation or GNB_P (290 nm), which was expected to have a short circulatory half-life due to its overall mass and size. $\alpha_v\beta_3$-targeted GNB_1 was produced by microfluidization as discussed before by incorporating an $\alpha_v\beta_3$-peptidomimetic antagonist. Neovascular imaging was performed in vivo in a Matrigel™ plug model of angiogenesis. For the first time, high spatial resolution noninvasive PAT imaging of angiogenesis was demonstrated using a 10 MHz ultrasound receiver that clearly revealed the formation of nascent neovessel tubules, sprouts, and bridges—much of which was still without blood flow. $\alpha_v\beta_3$-GNBM produced a 600% increase in signal in a Matrigel plug mouse model relative to the inherent hemoglobin contrast pretreatment. Competitive inhibition of $\alpha_v\beta_3$-GNBM with $\alpha_v\beta_3$-NB (no gold) blocked contrast enhancement to pretreatment levels. Similar images in the saline control animals showed no change in vascular anatomy over the same time course. Indeed, these images illustrate the genesis of neovasculature in the Matrigel plug model. Nontargeted GNB passively accumulated in the tortuous neovascularity, but provided low (less than half) contrast enhancement of the targeted agent. PA signal changes in the Matrigel plug were monitored serially over 5 hours or more (Figure 1.6).

Another approach to the issue of SLN PA imaging was the development of a sub-20 nm "soft" polymeric nanoparticle [1]. This agent was designed for rapid intraoperative administration with real-time PA imaging. The intraoperative approach helps to eliminate anatomic plane displacement, which can complicate detection between the preoperative images and patient repositioning in the operating room (OR). Moreover, by establishing a rapid OR procedure, the surgeon can determine the most direct approach to the node for resection, minimizing potential secondary complications of the axillary dissection. The obvious key to this procedure is rapid signal generation with adequate intraoperative persistence.

Early detection of immature, nascent angiogenic
vessels with photoacoustics imaging and GNB!

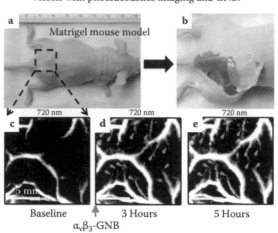

FIGURE 1.6   PA signal changes in the Matrigel™ plug were monitored serially over 5 hours
or more; (d) and (e) have arrows denoting small neovascular sprouts arising from immature
angiogenic vessels with only nascent blood flow.

1.5   CONCLUSIONS

The ability to monitor early and noninvasively in combination with targeted therapy
is a pressing need for emerging nanomedicine strategies. In this introductory chapter, we have discussed the opportunities and potential of targeted drug delivery and
molecular imaging with various imaging modalities for detection and treatment of
cardiovascular and related diseases with particular emphasis on US, MRI, and PAT.
The multidisciplinary field of nanomedicine is evolving fast and presenting new,
clinically relevant promises as the science of molecular biology, genomics, chemistry, and nanotechnology contributes greatly.

REFERENCES

1. R. Weissleder, U. Mahmood. 2001. Molecular imaging. *Radiology* (219):316.
2. T. F. Massoud, S. S. Gambhir. 2003. Molecular imaging in living subjects: Seeing fundamental biological processes in a new light. *Genes and Development* (17):545.
3. M. Doubrovin, I. Serganova, P. Mayer-Kuckuk, V. Ponomarev, R. G. Blasberg. 2004.
   Multimodality in vivo molecular-genetic imaging. *Bioconjugate Chemistry* (15):1376.
4. M.Thakur, B. C. Lentle. 2005. Report of a summit on molecular imaging. *Radiology*
   (236):753.
5. D. A. Mankoff. 2007. A definition of molecular imaging. *Journal of Nuclear Medicine*
   (48):18N.
6. (a) A. Lagaru, S. Chen, S. S. Gambhir. 2007. *Nature Clinical Practice Oncology*
   (4):556–557; (b) P. Richard. 2003. PET and the role of in vivo molecular imaging in
   personalized medicine. *Expert Review of Molecular Diagnostics* 3 (6):703–713.

7. (a) A. Gibson, J. Hebden, S. Arridge. 2005. Recent advances in diffuse optical imaging. *Physics in Medicine and Biology* (50):R1–R43; (b) S. Farsiu, J. Christofferson, B. Eriksson, P. Milanfar, B. Friedlander, A. Shakouri, R. Nowak. 2007. Statistical detection and imaging of objects hidden in turbid media using ballistic photons. *Applied Optics* (46):5805–5822; (c) B. Ballou et al. 1995. Tumor labeling in vivo using cyanine-conjugated monoclonal antibodies. *Cancer Immunology and Immunotherapy* (41):257–263; (d) R. Weissleder et al. 1999. A clearer vision for in vivo imaging. *Nature Biotechnology* (17):375–378; (e) P. Contag et al. 1998. Bioluminescent indicators in living mammals. *Nature Medicine* (4):245–247; (f) A. Becker et al. 2001. Receptor-targeted optical imaging of tumors with near-infrared fluorescent ligands. *Nature Biotechnology* (19):327–331; (g) L. A. Gross et al. 2000. The structure of the chromophore within DsRed, a red fluorescent protein from coral. *Proceedings of National Academy of Sciences USA* (97):11990–11995; (h) S. Achilefu et al. 2000. Novel receptor-targeted fluorescent contrast agents for in vivo tumor imaging. *Investigative Radiology* (35):479–485. (i) P. Alivisatos. 2004. The use of nanocrystals in biological detection. *Nature Biotechnology* (22):47–52; (j) F. F. Michalet, L. A. Pinaud, J. M. Bentolila, S. Tsay, J. J. Doose, G. Li, A. M. Sundaresan, et al. 2005. Quantum dots for live cells, in vivo imaging, and diagnostics. *Science* (307):538–544; (k) I. L. Medintz, H. T. Uyeda, E. R. Goldman, H. Mattoussi. 2005. Quantum dot bioconjugates for imaging, labeling and sensing. *Nature Materials* (4):435–446; (l) W. Cai, D. W. Shin, K. Chen, O. Gheysens, Q. Cao, S. X. Wang, S. S. Gambhir, et al. 2006. Peptide-labeled near-infrared quantum dots for imaging tumor vasculature in living subjects. *Nano Letters* (6):669–676; (m) B. Zhu, F. A. Jaffer, V. Ntziachristos, R. Weissleder. 2005. Development of a near infrared fluorescence catheter: Operating characteristics and feasibility for atherosclerotic plaque detection. *Journal of Physics D Applied Physics* 2701–2707; (n) X. Gao, L. Yang, J. A. Petros, F. F Marshall, J. W. Simons, S. Nie. 2005. *Current Opinion in Biotechnology* 16:63–72. (o) A. Smith, X. Gao, and S. Nie. 2004. Quantum-dot nanocrystals for in vivo molecular and cellular imaging. *Photochemistry and Photobiology* (80):377–385; (p) M. N. Rhyner, A. M. Smith, X. Gao, H. Mao, L. Yang, S. Nie. 2006. Quantum dots and multifunctional nanoparticles: New contrast agents for tumor imaging. *Nanomedicine* (1):209–217; (q) M. V. Yezhelyev, X. Gao, Y. Xing, A. Al-Hajj, S. Nie, Ruth M O'Regan. 2006. Emerging use of nanoparticles in diagnosis and treatment of breast cancer. *Lancet Oncology* (7):657–667; (r) M. Nahrendorf, D. E. Sosnovik, P. Waterman, F. K. Swirski, A. N. Pande, E. Aikawa, J. L. Figueiredo, et al. 2007. Dual channel optical tomographic imaging of leukocyte recruitment and protease activity in the healing myocardial infarct. *Circulation Research* (100):1218–1225.

8. H. Maeda, Y. Matsumura. 1989. Tumoritropic and lymphotropic principles of macromolecular drugs. *Critical Reviews in Therapeutic Drug Carrier Systems* (6):193–210.

9. N. Ohtsuka, T. Konno, Y. Miyauchi, H. Maeda. 1987. Anticancer effects of arterial administration of the anticancer agent SMANCS with lipiodol on metastatic lymph nodes. *Cancer* (59):1560–1565.

10. L. E. van Vlerken, T. K. Vyas, M. M. Amiji. 2007. Polyethyleneglycol-modified nanocarrier for tumor-targeted and intracellular delivery. *Pharmaceutical Research* (24):1405–1414.

11. (a) D. Shenoy, S. Little, R. Langer, M. Amiji. 2005. Poly(ethylene oxide)-modified poly(beta-amino ester) nanoparticles as a pH-sensitive system for tumor-targeted delivery of hydrophobic drugs: Part 2. In vivo distribution and tumor localization studies. *Pharmaceutical Research* (22):2107–2114; (b) D. Pan, N. Sanyal, A. H. Schmieder, A. Senpan, B. Kim, X. Yang, G. Hu, J. S. Allen, R. W. Gross, S. A. Wickline, G. M. Lanza. 2012. *Nanomedicine* (London). (10):1507–1519.

12. D. Shenoy, S. Little, R. Langer, M. Amiji. 2005. Poly(ethylene oxide)-modified poly(beta-amino ester) nanoparticles as a pH-sensitive system for tumor-targeted delivery of hydrophobic drugs. 1. In vitro evaluations. *Molecular Pharmacology* (2):357–366.

13. D. B. Shenoy, M. M. Amiji. 2005. Poly(ethylene oxide)-modified poly(epsilon-caprolactone) nanoparticles for targeted delivery of tamoxifen in breast cancer. *International Journal of Pharmaceutics* (293):261–270.

14. H. Devalapally, Z. Duan, M. V. Seiden, M. M. Amiji. 2007. Paclitaxel and ceramide coadministration in biodegradable polymeric nanoparticulate delivery system to overcome drug resistance in ovarian cancer. *International Journal of Cancer* (121):1830–1838.

15. L. E. van Vlerken, Z. Duan, M. V. Seiden, M. M. Amiji. 2007. Modulation of intracellular ceramide using polymeric nanoparticles to overcome multidrug resistance in cancer. *Cancer Research* (67):4843–4850.

16. D. Patrick O'Neala, L. R. Hirschb, N. J. Halasc, J. D. Paynea, J. L. West. 2004. Photothermal tumor ablation in mice using near infrared-absorbing nanoparticles. *Cancer Letters* (209):171–176.

17. S. H. Britz-Cunningham, S. J. Adelstein. 2003. Molecular targeting with radionuclides: State of the science. *Journal of Nuclear Medicine* (44):1945–1961.

18. S. F. Rosebrough, M. Hashmi. 1996. Galactose-modified streptavidin-GC4 antifibrin monoclonal antibody conjugates: Application for two-step thrombus/embolus imaging. *Journal of Pharmacology and Experimental Therapeutics* (276):770–775.

19. H. Tada, H. Higuchi, T. M. Watanabe, N. Ohuchi. 2007. In vivo real-time tracking of single quantum dots conjugated with monoclonal anti-HER2 antibody in tumors of mice. *Cancer Research* (67):1138–1144.

20. R. Pasqualini, E. Koivunen, E. Ruoslahti. 1997. Alpha v integrins as receptors for tumor targeting by circulating ligands. *Nature Biotechnology* (15):542.

21. D. Pan, J. L. Turner, K. L. Wooley. 2003. Folic acid-conjugated nanostructured materials designed for cancer cell targeting. *Chemistry Communications* 2400–2401.

22. R. Rossin, D. Pan, K. Qi, J. L. Turner, X. Sun, K. L. Wooley, M. J. Welch. 2004. ^{64}Cu-labeled folate-conjugated shell cross-linked nanoparticles for tumor imaging and radiotherapy: Synthesis, radiolabeling, and biologic evaluation. *Journal of Nuclear Medicine* (46):1210–1218.

23. X. Y. Chen, Y. Hou, M. Tohme, R. Park, V. Khankaldyyan, I. Gonzales-Gomez, J. R. Bading, W. E. Laug, et al. 2004. PEGylated Arg-Gly-Asp peptide: [64]Cu labeling and PET imaging of brain tumor alphavbeta3-integrin expression. *Journal of Nuclear Medicine* (45):1776.

24. J. D. Hood, M. Bednarski, R. Frausto, S. Guccione, R. A. Reisfeld, R. Xiang, R., D. A. Cheresh. 2002. Tumor regression by targeted gene delivery to the neovasculature. *Science* (296):2404.

25. D. Pan, J. L.Turner, K. L. Wooley. 2004. Shell cross-linked nanoparticles designed to target angiogenic blood vessels via $\alpha_v\beta_3$ receptor–ligand interactions. *Macromolecules* (37):7109–7115.

26. E. N. Brody, L. Gold. 2000. Aptamers as therapeutics and diagnostic agents. *Journal of Biotechnology* (74):5–13.

27. S. Missailidis, A. Perkins. 2007. Aptamers as novel radiopharmaceuticals: Their applications and future prospects in diagnosis and therapy. *Cancer Biotherapy & Radiopharmaceuticals* (22):453–468.

28. O. C. Farokhzad, J. M. Karp, R. Langer. 2006. Nanoparticle–aptamer bioconjugates for cancer targeting. *Expert Opinions in Drug Delivery* (3):311–324.

29. A. Maruyama, T. Ishihara, J.-S. Kim, S. W. Kim, T. Akaike. 1997. Nanoparticle DNA carrier with poly(L-lysine) grafted polysaccharide copolymer and poly(D,L-lactic acid). *Bioconjugate Chemistry* (8):735–742.

30. R. Mehvar. 2003. Recent trends in the use of polysaccharides for improved delivery of therapeutic agents: Pharmacokinetic and pharmacodynamic perspectives. *Current Pharmaceutical Biotechnology* (4):283–302.

31. R. Gref, Y. Minamitake, M. T. Peracchia, V. Trubetskoy, V. Torchilin, R. Langer. 1994. Biodegradable long-circulating polymeric nanospheres. *Science* (263):1600–1603.

32. M. A. Mazid, E. Moase, E. Scott, H. R. Hanna, F. M. Unger. 1991. Synthesis and bioactivity of copolymers with fragments of heparin. *Journal of Biomedicine and Materials Research* (25):1169–1181.

33. S. Saini, R. R. Edelman, P. Sharma, et al. 1995. Blood-pool MR contrast material for detection and characterization of focal hepatic lesions: Initial clinical experience with ultrasmall superparamagnetic iron oxide (AMI-227). *American Journal of Roentgenology* (164):1147–1152.

34. M. G. Harisinghani, J. Barentsz, P. F. Hahn, W. M. Deserno, S. Tabatabaei, C. H. van de Kaa, J. de la Rosette, et al. 2003. Noninvasive detection of clinically occult lymph-node metastases in prostate cancer. *New England Journal of Medicine* (348):2491–2499.

35. S. G. Ruehm, C. Corot, P. Vogt, S. Kolb, J. F. Debatin. 2001. Magnetic resonance imaging of atherosclerotic plaque with ultrasmall superparamagnetic particles of iron oxide in hyperlipidemic rabbits. *Circulation* (103) 415–422.

36. A. Quintana, E. Raczka, L. Piehler, et al. 2002. Design and function of a dendrimer-based therapeutic nanodevice targeted to tumor cells through the folate receptor. *Pharmaceutical Research* (19):1310–1316.

37. H. Kobayashi, M. W. Brechbiel. 2003. Dendrimer-based macromolecular MRI contrast agents: Characteristics and application. *Molecular Imaging* (2):1–10.

38. J. L. Turner, D. Pan, et al. 2005. Synthesis of gadolinium-labeled shell-cross-linked nanoparticles for magnetic resonance imaging applications. *Advanced Functional Materials* (15):1248–1254.

39. A. Agrawal, S. Huang, A. Wei Haw Lin, M. H. Lee, J. K. Barton, R. A. Drezek, T. J. Pfefer. 2006. Quantitative evaluation of optical coherence tomography signal enhancement with gold nanoshells. *Journal of Biomedical Optics* (11):041121.

40. C. Loo, A. Lin, L. Hirsch, M. H. Lee, J. Barton, N. Halas, J. West, R. Drezek. 2004. Nanoshell-enabled photonics-based imaging and therapy of cancer. *Technology in Cancer Research Treatment* (3):33–40.

41. H. Cang, T. Sun, Z. Y. Li, J. Chen, B. J. Wiley, Y. Xia, X. Li. 2005. Gold nanocages as contrast agents for spectroscopic optical coherence tomography. *Optics Letters* (30):3048–3050.

42. K. Byeong-Su, Q. Jiao-Ming, W. Jian-Ping, T. A. Taton. 2005. Magnetomicelles: Composite nanostructures from magnetic nanoparticles and cross-linked amphiphilic block copolymers. *Nano Letters* (5):1987–1991.

43. Q. Wang, T. Lin, L. Tang, J. E. Johnson, M. G. Finn. 2002. Icosahedral virus particles as addressable nanoscale building blocks. *Angewandt Chemie* (International Edition) (41):459–462.

44. M. Manchester, P. Singh. 2006. Virus-based nanoparticles (VNPs): Platform technologies for diagnostic imaging. *Advanced Drug Delivery Reviews* (58):1505.

45. L. Josephson, C. H. Tung, A. Moore, R. Weissleder. 1999. High-efficiency intracellular magnetic labeling with novel superparamagnetic-tat peptide conjugates. *Bioconjugate Chemistry* (10):186–191.

46. C. H. Dodd, H. C. Hsu, H, W. J. Chu, et al. 2001. Normal T-cell response and in vivo magnetic resonance imaging of T cells loaded with HIV transactivator-peptide-derived superparamagnetic nanoparticles. *Journal of Immunology Methods* (256):89–105.

47. H. W. Kang, L. Josephson, A. Petrovsky, R. Weissleder, A. Bogdanov, Jr. 2002. Magnetic resonance imaging of inducible E-selectin expression in human endothelial cell culture. *Bioconjugate Chemistry* (13):122–127.

48. E. A. Schellenberger, A. Bogdanov, Jr., D. Hogemann, J. Tait, R. Weissleder, L. Josephson. 2002. Annexin V-CLIO: A nanoparticle for detecting apoptosis by MRI. *Molecular Imaging* (1):102–107.

49. D. Artemov, N. Mori, B. Okollie, Z. M. Bhujwalla. 2003. MR molecular imaging of the Her-2/neu receptor in breast cancer cells using targeted iron oxide nanoparticles. *Magnetic Resonance Medicine* (49):403–408.

50. B. A. Moffat, G. R. Reddy, P. McConville, et al. 2003. A novel polyacrylamide magnetic nanoparticle contrast agent for molecular imaging using MRI. *Molecular Imaging* (2):324–332.

51. H. Kobayashi, S. Kawamoto, S. K. Jo, et al. 2002. Renal tubular damage detected by dynamic micro-MRI with a dendrimer-based magnetic resonance contrast agent. *Kidney International* (61):1980–1985.

52. H. Kobayashi, S. Kawamoto, M. W. Brechbiel, et al. 2004. Micro-MRI methods to detect renal cysts in mice. *Kidney International* 65:1511–1516.

53. J. W. Dear et al. 2005. Dendrimer-enhanced MRI as a diagnostic and prognostic biomarker of sepsis-induced acute renal failure in aged mice. *Kidney International* (67):2159–2167.

54. K. Hisataka, M. W. Brechbiel. 2005. Nano-sized MRI contrast agents with dendrimer cores. *Advanced Drug Delivery Reviews* (57):2271–2286.

55. S. E. Stiriba, H. Frey, R. Haag. 2002. Dendritic polymers in biomedical applications: From potential to clinical use in diagnostics and therapy. *Angewandte Chemie International Edition* (41):1329–1334.

56. H. Kobayashi, S. Kawamoto, T. Saga, N. Sato, T. Ishimori, J. Konishi, K. Ono, et al. 2001. Avidin dendrimer-(1B4M-Gd)(254). *Bioconjugate Chemistry* (12):587–593.

57. C. S. Winalski, S. Shortkroff, E. Schneider, H. Yoshioka, R. V. Mulkern, G. M. Rosen. 2008. Targeted dendrimer-based contrast agents for articular cartilage assessment by MR imaging. *Osteoarthritis Cartilage* (16):815–822.

58. G. Quido, L. Sander, G. H. Regina, H. P. Marcel, A. W. Griffioen, J. M. A. van Engelshoven, W. H. Backes. 2005. Dynamic contrast-enhanced MR imaging kinetic parameters and molecular weight of dendritic contrast agents in tumor angiogenesis in mice. *Radiology* (235):65–72.

59. C. E. Neumaier, G. Baio, S. Ferrini, G. Corte, A. Daga. 2008. MR and iron magnetic nanoparticles. Imaging opportunities in preclinical and translational research. *Tumori* (94):226–233.

60. (a) G. Navon, R. Panigel, G. Valensin. 1986. *Magnetic Resonance Medicine* (3):876–880; (b) E. C. Unger, T. Porter, W. Culp, R. Labell, T. Matsunaga, R. Zutshi. 2004. *Advances in Drug Delivery Reviews* (56):1291–1314; (c) K. C. Li, M. D. Bednarski. 2004. *Journal of Magnetic Resonance Imaging* (16):388–393.

61. S. H. Koenig, Q. F. Ahkong, R. D. Brown, et al. 1992. *Magnetic Resonance Medicine* (23):275–286.

62. C. W. Grant, S. Karlik, E. Florio. 1989. *Magnetic Resonance Medicine* (11):236–243.

63. G. Kabalka, M. Davis, E. Holmberg, K. Maruyama, L. Huang. 1991. *Magnetic Resonance Medicine* (9):373–377.

64. D. A. Sipkins, D. A. Cheresh, M. R. Kazemi, L. M. Nevin, M. D. Bednarski, K. C. Li. 1998. Detection of tumor angiogenesis in vivo by v3-targeted magnetic resonance imaging. *Nature Medicine* (4):623–626.

65. J. D. Hood, D. A. Cheresh. 2002. Targeted delivery of mutant Raf kinase to neovessels causes tumor regression. *Cold Spring Harbor Symposium* (67): 285–291.

66. S. Fossheim, A. Fahlvik, J. Klaveness, R. Muller. 1999. *Magnetic Resonance Medicine* (17):83–89.

67. B. A. Moffat, G. R. Reddy, P. McConville, et al. 2003. A novel polyacrylamide magnetic nanoparticle contrast agent for molecular imaging using MRI. *Molecular Imaging* (2):324–332.

68. (a) G. M. Lanza, K. D. Wallace, M. J. Scott, et al. 1996. A novel site-targeted ultrasonic contrast agent with broad biomedical application. *Circulation* (94):3334–3340; (b) P. M. Winter, A. H. Schmieder, S. D. Caruthers, J. L. Keene, H. Zhang, S. A. Wickline, G. M. Lanza. 2008. Minute dosages of β3-targeted fumagillin nanoparticles impair Vx-2 tumor angiogenesis and development in rabbits. *FASEB Journal* (22):2758–2767; (c) A. H. Schmieder, P. M. Winter, S. D. Caruthers, T. D. Harris, T. A. Williams, J. S. Allen, E. K. Lacy, et al. 2005. Molecular MR imaging of melanoma angiogenesis with alphanubeta3-targeted paramagnetic nanoparticles. *Magnetic Resonance Medicine* (53):621–627; (d) P. M. Winter, S. D. Caruthers, A. Kassner, T. D. Harris, L. K. Chinen, J. S. Allen, E. K. Lacy, et al. 2005. Molecular imaging of angiogenesis in nascent Vx-2 rabbit tumors using a novel alpha(nu)beta3-targeted nanoparticle and 1.5 tesla magnetic resonance imaging. *Cancer Research* (63):5838–5843.

69. S. A. Anderson, R. K. Rader, W. F. Westlin, C. Null, D. Jackson, G. M. Lanza, S. A. Wickline, et al. 2000. Magnetic resonance contrast enhancement of neovasculature with avb3-targeted nanoparticles. *Magnetic Resonance Medicine* (44):433–439.

70. D. A. Sipkins, D. A. Cheresh, M. R. Kazemi, L. M. Nevin, M. D. Bednarski, K. C. Li. 1998. Detection of tumor angiogenesis in vivo by alphaVbeta3-targeted magnetic resonance imaging. *Nature Medicine* (4):623–626.

71. G. T. Hermanson. 2008. *Bioconjugate techniques*, 2nd ed. London: Academic Press.

72. M. G. Finn, K. B. Sharpless, et. al. 2003. Bioconjugation by copper(I)-catalyzed azide-alkyne [3 + 2] cycloaddition. *Journal of American Chemical Society* (125):3192.

73. D. Chandrasekar D. et al. 2007. The development of folate-PAMAM dendrimer conjugates for targeted delivery of anti-arthritic drugs and their pharmacokinetics and biodistribution in arthritic rats. *Biomaterials* (28):504–512.

74. N. Fischer-Durand et al. 2007. *Macromolecules* (40):8568–8575.

75. L. Wang et al. 1998. *Bioconjugate Chemistry* (9):749–757.

76. T. Heinze, T. Liebert, B. Heublein, S. Hornig. 2006. Functional polymers based on dextran: Book series advances in polymer. *Science* (205):199–291.

77. H. C. Kolb, M. G. Finn, K. B. Sharpless. 2001. Click chemistry: Diverse chemical function from a few good reactions. *Angewandte Chemie* International Edition (40):2004–2021.

78. A. J. Link, D. A. Tirrell. 2003. Cell surface labeling of *Escherichia coli* via copper(i)-catalyzed [3+2] cycloaddition. *Journal of American Chemical Society* (125):11164–11165.

79. Q. Wang, T. Lin, L. Tang, J. E. Johnson, M. G. Finn. 2002. Icosahedral virus particles as addressable nanoscale building blocks. *Angewandte Chemie* International Edition (41):459–462.

80. S. S. Gupta, J. Kuzelka, P. Singh, W. G. Lewis, M. Manchester, M. G. Finn. 2005. Accelerated bioorthogonal conjugation: A practical method for the ligation of diverse functional molecules to a polyvalent virus scaffold. *Bioconjugate Chemistry* (16):1572–1579.

81. A. J. Dirks, S. S. Van Berkel, N. S. Hatzakis, J. A. Opsteen, F. L. Van Delft, J. J. L. M. Cornelissen, A. E. Rowan, J. C. M. Van Hest, F. P. J. T. Rutjes, R. J. M. Nolte. 2005. *Chemistry Communications* 4172–4174.

82. A. J. Dirks, R. J. M. Nolte, J. J. L. M. Cornelissen. 2008. Protein-polymer hybrid amphiphiles, *Advanced Materials*, 20(20):3953–3957.

83. N. K. Devaraj, G. P. Miller, W. Ebina, B. Kakaradov, J. P. Collman, E. T. Kool, C. E. D. Chidsey. 2005. Chemoselective covalent coupling of oligonucleotide probes to self-assembled monolayers. *Journal of American Chemical Society* (127):127, 8600–8601.

84. H. Jang, A. Fafarman, J. M. Holub, K. Kirshenbaum. 2005. *Organic Letters* (7):1951–1954.
85. X. L. Sun, C. L. Stabler, C. S. Cazalis, E. L. Chaikof. 2006. *Bioconjugate Chemistry* (17):52–57.
86. H-D. Liang, M. J. K. Blomley. 2003. *British Journal of Radiology* (76):S140–S150.
87. S. H. Bloch, P. A. Dayton, K. W. Ferrara. 2004. *IEEE Engineering in Medicine and Biology Magazine* 23:18–29.
88. T. F. Massoud, S. S. Gambhir. 2003. *Genes Development* (17):545–580.
89. D. Cosgrove. 2006. *European Journal of Radiology* (60):324–330.
90. J. A. Jakobsen. 2001. *European Journal of Dermatology and European Radiology* (11):1329–1337.
91. E. Stride, N. Saffari. 2003. *Proceedings of Institution of Mechanical Engineers Part H* (217):429–447.
92. D. B. Ellegala, H. Leong-Poi, J. E. Carpenter, A. L. Klibanov, S. Kaul, M. E. Shaffrey, J. Sklenar, J. R. Lindner. 2003. *Circulation* (108):336–341.
93. H. Leong-Poi, J. Christiansen, A. L. Klibanov, S. Kaul, J. R. Lindner. 2003. *Circulation* (107):455–460.
94. H. Leong-Poi, J. Christiansen, P. Heppner, C. W. Lewis, A. L. Klibanov, S. Kaul, J. R. Lindner. 2005. *Circulation* (111):3248–3254.
95. G. Korpanty, J. G. Carbon, P. A. Grayburn, J. B. Fleming, R. A. Brekken. 2007. *Clinical Cancer Research* (13):323–330.
96. J. K. Willmann et al. 2008. *Radiology* (246):508–518.
97. G. Korpanty, J. G. Carbon, P. A. Grayburn, J. B. Fleming, R. A. Brekken. 2007. *Clinical Cancer Research* (13) 323–330.
98. M. S. Hughes, J. N. Marsh, et al. 2005. *Journal of Acoustical Society of America* (117):964–972.
99. G. M. Lanza, R. L. Trousil, et al. 1998. *Journal of Acoustical Society of America* (104):3665–3672.
100. G. M. Lanza, D. R. Abendschein, C. S. Hall, M. J. Scott, D. E. Scherrer, A. Houseman, J. G. Miller, et al. 2000. *Journal of American Society of Echocardiography* (13):608–614.
101. G. M. Lanza, D. R. Abendschein, C. S. Hall, J. N. Marsh, M. J. Scott, D. E. Scherrer, S. A. Wickline. 2000. *Investigative Radiology* (35):227–234.
102. M. S. Hughes, J. N. Marsh, H. Zhang, A. K. Woodson, J. S. Allen, E. K. Lacy, C. Carradine, et al. 2006. *IEEE Transactions of Ultrasonics Ferroelectrics and Frequency Control* (53):1609–1616.
103. S. Lavisse, R. A. Paci, C. Adotevi, P. Opolon, P. Peronneau, P. Bourget, A. Roche, et al. 2005. In vitro echogenicity characterization of poly[lactide-coglycolide] (PLGA) microparticles and preliminary in vivo ultrasound enhancement study for ultrasound contrast agent application. *Investigative Radiology* (40):536–544.
104. H. E. Longmaid, III, D. F. Adams et al. 1971. *Investigative Radiology* (20):141–145.
105. R. Damadian. 1971. *Science* (171):1151–1153.
106. A. de Roos, J. Doornbos, D. Baleriaux, H. L. Bloem, T. H. Falke. 1988. In *Magnetic resonance annual,* ed. H. Y. Kressel, 113–145. New York: Raven Press.
107. Martin R. Prince, E. Kent Yucel, John A. Kaufman, David C. Harrison, Stuart C. Geller. 1993. Dynamic gadolinium-enhanced three-dimensional abdominal MR arteriography. *Journal of Magnetic Resonance Imaging* (3):877–881.
108. J. A. Kaufman, S. C. Geller, A. C. Waltman. 1998. Renal insufficiency: Gadopentetate dimeglumine as a radiographic contrast agent during peripheral vascular interventional procedures. *Radiology* (198):579–581.

109. (a) D. L. Thorek, A. K. Chen, J. Czupryna, A. Tsourkas. 2005. *Annals of Biomedical Engineering* 34:23–38; (b) Z. Zhang, S. A. Nair, T. J. McMurry. 2005. *Current Medical Chemistry* (12):751–778.

110. G. M. Lanza, P. M. Winter, et al. 2004. *Journal of Nuclear Cardiology* (11):733–743.

111. K. L. Nelson, V. M. Runge. 1995. Basic principles of MR contrast. *Topics in Magnetic Resonance Imaging* (7):124–136.

112. M. D. Chavanpatil, A. Khdair, J. Panyam. 2006. *Journal of Nanoscience and Nanotechnology* (6):2651–2663.

113. K. Shamsi, T. Balzer, S. Saini, P. R. Ros, R. C. Nelson, E. C. Carter, S. Tollerfield, et al. 1998. *Radiology* (206):365–371.

114. P. Reimer, N. Jahnke, M. Fiebich, W. Schima, F. Deckers, C. Marx, N. Holzknecht, et al. 2000. *Radiology* (217):152–158.

115. R. Weissleder, P. F. Hahn, D. D. Stark, G. Elizondo, S. Saini, L. E. Todd, J. Wittenberg, et al. 1988. *Radiology* (169):399–403.

116. R. Weissleder, D. D. Stark, E. J. Rummeny, C. C. Compton, J. T. Ferrucci. 1988. *Radiology* (166):423–430.

117. M. G. Mack, J. O. Balzer, R. Straub, K. Eichler, T. J. Vogl. 2002. *Radiology* (222):239–244.

118. Y. Anzai, C. W. Piccoli, E. K. Outwater, W. Stanford, D. A. Bluemke, P. Nurenberg, S. Saini, et al. 2003. *Radiology* (228):777–788.

119. W. S. Enochs, G. Harsh, F. Hochberg, R. Weissleder. 1999. *Journal of Magnetic Resonance Imaging* (9):228–232.

120. A. Saleh, M. Schroeter, C. Jonkmanns, H. P. Hartung, U. Modder, S. Jander. 2004. *Brain* (127):1670–1677.

121. C. Zimmer, R. Weissleder, K. Poss, A. Bogdanova, S. C. Wright, Jr., W. S. Enochs. 1995. *Radiology* (197):533–538.

122. A. Moore, E. Marecos, A. Bogdanov, Jr., R. Weissleder. 2000. *Radiology* (214):568–574.

123. M. G. Harisinghani, J. Barentsz, P. F. Hahn, W. M. Deserno, S. Tabatabaei, C. H. van de Kaa, J. de La Rosette, et al. 2003. *New England Journal of Medicine* (348):2491–2499.

124. V. Dousset, B. Brochet, M. S. Deloire, L. Lagoarde, B. Barroso, J. M. Caille, K. G. Petry. 2006. *American Journal of Neuroradiology* (27):1000–1005.

125. M. E. Kooi, V. C. Cappendijk, et al. 2003. Accumulation of ultrasmall superparamagnetic particles of iron oxide in human atherosclerotic plaques can be detected by in vivo magnetic resonance imaging. *Circulation* (107):2453–2458.

126. C. Corot, K. G. Petry, R. Trivedi, A. Saleh, C. Jonkmanns, J. F. Le Bas, E. Blezer, et al. 2004. *Investigative Radiology* (39):619–625.

127. D. J. Stuckey, C. A. Carr, E. Martin-Rendon, D. J. Tyler, C. Willmott, P. J. Cassidy, S. J. Hale, et al. 2006. *Stem Cells* (24):1968–1975.

128. A. S. Arbab, E. K. Jordan, L. B. Wilson, G. T. Yocum, B. K. Lewis, J. A. Frank. 2004. *Human Gene Therapy* (15):351–360.

129. J. M. Hill, A. J. Dick, V. K. Raman, R. B. Thompson, Z. X. Yu, K. A. Hinds, B. S. Pessanha, et al. 2003. *Circulation* (108):1009–1014.

130. I. J. de Vries, W. J. Lesterhuis, J. O. Barentsz, P. Verdijk, J. H. van Krieken, O. C. Boerman, W. J. Oyen, et al. 2005. *Nature Biotechnology* (23):1407–1413.

131. E. T. Ahrens, M. Feili-Hariri, H. Xu, G. Genove, P. A. Morel. 2003. *Magnetic Resonance Medicine* (49):1006–1013.

132. E. M. Shapiro, S. Skrtic, K. Sharer, J. M. Hill, C. E. Dunbar, A. P. Koretsky. 2004. *Proceedings of National Academy of Sciences USA* (101):10901–10906.

133. P. Caravan, J. J. Ellison, T. J. McMurry, R. B. Lauffer. 1999. *Chemical Reviews* (99):2293.

134. A. M. Ponce et al. 2007. *Journal of National Cancer Institute* (99):53–63.

135. R. A. Schwendener et al. 1989. *International Journal of Pharmaceutics* (3):249–259.

136. J. H. Lee, A. P. Koretsky. 2004. Manganese enhanced magnetic resonance imaging. *Current Pharmaceutical Biotechnology* (5):529–537.
137. A. P. Koretsky, A. C. Silva. 2004. Manganese-enhanced magnetic resonance imaging (MEMRI). *NMR Biomedicine* (17):527–531.
138. R. G. Pautler. 2006. Biological applications of manganese-enhanced magnetic resonance imaging. *Methods in Molecular Medicine* (124):365–386.
139. N. A. Bock, A. C. Silva. 2007. Manganese: A unique neuroimaging contrast agent. *Future Neurology* (2):297–395.
140. S. Flacke, S. Fischer, M. J. Scott, R. J. Fuhrhop, J. S. Allen, M. McLean, P. Winter, et al. 2001. Novel MRI contrast agent for molecular imaging of fibrin: Implications for detecting vulnerable plaques. *Circulation* (104):1280–1285.
141. P. M. Winter, S. D. Caruthers, et al. 2003. *Magnetic Resonance in Medicine* (50):411–416.
142. G. J. Stanisz, R. M. Henkelman. 2000. *Magnetic Resonance in Medicine* (44):665–667.
143. A. M. Morawski, P. M. Winter, et al. 2004. *Magnetic Resonance in Medicine* (52):1255–1262.
144. (a) R. B. Lauffer, D. J. Parmelee, S. U. Dunham, H. S. Ouellet, R. P. Dolan, S. Witte, T. J. McMurry et al. 1998. *Radiology* (207):529–538; (b) D. J. Parmelee, R. C. Walovitch, H. S. Ouellet, R. B. Lauffer. 1997. *Investigative Radiology* (32):741–747.
145. P. H. Kuo. 2008. *Journal of American Collegeof Radiology* (5):29–35.
146. A. K. Abu-Alfa. 2008. *Journal of American College of Radiology* (5):45–52.
147. H. Ersoy, F. J. Rybicki. 2007. *Journal of Magnetic Resonance Imaging* (26):1190–1197.
148. H. B. Na, J. H. Lee, et al. 2007. *Angewandte Chemie* International Edition (46):5397–5401.
149. D. Pan, S. D. Caruthers, et al. 2008. *Journal of American Chemical Society* (130):9186.
150. (a) A. Y. Louie, M.M. Huber, E. T. Ahrens, U. Rothbacher, R. Moats, R. E. Jacobs, S.E. Fraser, T. J. Meade. 2000. In vivo visualization of gene expression using magnetic resonance imaging. *Nature Biotechnology* (18):321–325; (b) R. E. Jacobs, E. T. Ahrens, T. J. Meade, S. E. Fraser. 1999. Looking deeper into vertebrate development. *Trends Cell Biology* (9):73–76. (c) E. M. Shapiro, A. P. Koretsky. 2008. *Magnetic Resonance in Medicine* (60):265–269.
151. R. Benson. 1926. Present status of coronary artery disease. *Archives Pathology Lab Medicine* (2):876–916.
152. P. Constantinides. 1966. Plaque fissures in human coronary thrombosis. *Journal of Atherosclerotic Research* (6):1–17.
153. R. Virmani, A. P. Burke, A. Farb, F. D. Kolodgie. 2006. Pathology of the vulnerable plaque. *Journal of American College of Cardiology* (47):(8 SUPPL.)
154. R. Virmani, E. R. Ladich, A. P. Burke, et al. 2006. Histopathology of carotid atherosclerotic disease. *Neurosurgery* (59):(5 Suppl 3):S219–227.
155. F. D. Kolodgie, R. Virmani, A. P. Burke, A. Farb, D. K. Weber, R. Kutys, A. V. Finn, H. K. Gold. 2004. Pathologic assessment of the vulnerable human coronary plaque. *Heart* (90):1385–1391.
156. J. A. Schaar, J. E. Muller, E. Falk, R. Virmani, V. Fuster, P. W. Serruys, A. Colombo, C. Stefanadis, S. W. Casscells, R. R. Moreno, A. Maseri, A. F. W. Van Der Steen. 2004. Terminology for high-risk and vulnerable coronary artery plaques. *European Heart Journal* (25):1077–1082.
157. H. Huang, R. Virmani, H. Younis, et al. 2001. The impact of calcification on the biomechanical stability of atherosclerotic plaques. *Circulation* (103):1051–1056.
158. Z. H. Saskia, M. D. Rittersma, C. Allard van der Wal, et al. 2005. Plaque instability frequently occurs days or weeks before occlusive coronary thrombosis. *Circulation* (111):1160–1165.

159. S. Ojio, H. Takatsu, T. Tanaka, K. Ueno, K. Yokoya, T. Matsubara, T. Suzuki, et al. 2000. Considerable time from the onset of plaque rupture and/or thrombi until the onset of acute myocardial infarction in humans: Coronary angiographic findings within 1 week before the onset of infarction. *Circulation* (102):2063–2069.

160. J. Mann, M. J. Davies. 1999. Mechanisms of progression in native coronary artery disease: Role of healed plaque disruption. *Heart* (82):265–268.

161. J. Mann, M. J. Davies. 1995. Assessment of the severity of coronary artery disease at postmortem examination. Are the measurements clinically valid? *British Heart Journal* (74):528–530.

162. A. R. Moody, S. Allder, G. Lennox et al. 1999. Direct magnetic resonance imaging of carotid artery thrombus in acute stroke. *Lancet* (353):122–123.

163. K. Nasu, E. Tsuchikane, O. Katoh, D. G. Vince, R. Virmani, J. F. Surmely, A. Murata, et al. 2006. Accuracy of in vivo coronary plaque morphology assessment. A validation study of in vivo virtual histology compared with in vitro histopathology. *Journal of American College of Cardiology* 47 (12):2405–2412.

164. I. K. Jang, G. J. Tearney, B. MacNeill, M. Takano, F. Moselewski, N. Iftima, M. Shishkov, et al. 2005. In vivo characterization of coronary atherosclerotic plaque by use of optical coherence tomography. *Circulation* 111 (12):1551–1555.

165. F. Ishibashi, K. Aziz, G. S. Abela, S. Waxman. 2006. Update on coronary angioscopy: Review of a 20-year experience and potential application for detection of vulnerable plaque. *Journal of Interventional Cardiology* 19 (1):17–25.

166. M. Skinner, C. Yuan, L. Mitsumori, et al. 1995. Serial magnetic resonance imaging of experimental atherosclerosis detects lesion fine structure, progression and complications in vivo. *Nature Medicine* (1):69–73.

167. J. Toussaint, G. LaMuraglia, J. Southern, et al. 1996. Magnetic resonance images lipid, fibrous, calcified, hemorrhagic, and thrombotic components of human atherosclerosis in vivo. *Circulation* (94):932–938.

168. C. Yuan, J. S. Tsuruda, K. N. Beach, C. E. Hayes, M. S. Ferguson, C. E. Alpers, T. K. Foo, et al. Techniques for high-resolution MR imaging of atherosclerotic plaque. *Journal of Magnetic Resonance Imaging* 4(1):43–49.

169. J-F. Toussaint, G. M. LaMuraglia, J. F. Southern, V. Fuster, H. L. Kantor. 1996. Magnetic resonance images lipid, fibrous, calcified, hemorrhagic, and thrombotic components of human atherosclerosis in vivo. *Circulation* (94):932–938.

170. T. Hatsukami, R. Ross, N. Polissar, et al. 2000. Visualization of fibrous cap thickness and rupture in human atherosclerotic carotid plaque in vivo with high-resolution magnetic resonance imaging. *Circulation* (102):959–964.

171. Z. Fayad, V. Fuster. 2000. Characterization of atherosclerotic plaques by magnetic resonance imaging. *Annals New York Academy Sciences* (902):173–186.

172. Z. Fayad, V. Fuster, J. Fallon, et al. 2000. Noninvasive in vivo human coronary artery lumen and wall imaging using black-blood magnetic resonance imaging. *Circulation* (102) 506–510.

173. A. P. King, R. Boubertakh, K. L. Ng, Y. L. Ma, P. Chinchapatnam, G. Gao, T. Schaeffter, et al. 2008. A technique for respiratory motion correction in image guided cardiac catheterisation procedures. Paper presented at *Progress in Biomedical Optics and Imaging— Proceedings of SPIE.*

174. F. Odille, N. Cindea, D. Mandry, C. Pasquier, P. A. Vuissoz, J. Felblinger. 2008. Generalized MRI reconstruction including elastic physiological motion and coil sensitivity encoding. *Magnetic Resonance in Medicine* 59 (6): 1401–1411.

175. J. R. Maclaren, P. J. Bones, R. P. Millane, R. Watts. 2008. MRI with TRELLIS: A novel approach to motion correction. *Magnetic Resonance Imaging* 26 (4): 474–483.

176. S. Flacke, S. Fischer, C. Hall, et al. 1999. Targeted magnetic resonance contrast agent for detection of thrombus. *Journal of Cardiovascular Magnetic Resonance* (1):353.

177. R. M. Botnar, A. S. Perez, S. Witte, A. J. Wiethoff, et al. 2004. In vivo molecular imaging of acute and subacute thrombosis using a fibrin-binding magnetic resonance imaging contrast agent. *Circulation* (109):2023–2029.

178. X. Yu, S. K. Song, J. Chen, et al. 2000. High-resolution MRI characterization of human thrombus using a novel fibrin-targeted paramagnetic nanoparticle contrast agent. *Magnetic Resonance Medicine* 4 (6):867–872.

179. A. M. Morawski, P. M. Winter, K. C. Crowder, et al. 2004. Targeted nanoparticles for quantitative imaging of sparse molecular epitopes with MRI. *Magnetic Resonance Medicine* 51 (3): 480–486.

180. P. M. Winter, S. D. Caruthers, X. Yu, et al. 2003. Improved molecular imaging contrast agent for detection of human thrombus. *Magnetic Resonance Medicine* 50 (2): 411–416.

181. P. M. Winter, P. Athey, G. Kiefer, G. Gulyas, K. Frank, R. Fuhrhop, D. Robertson, et al. 2005. Improved paramagnetic chelate for molecular imaging with MRI. *Journal of Magnetic Materials* (293):540–545.

182. A. M. Neubauer, S. D. Caruthers et al. 2007. *Journal of Cardiovascular Magnetic Resonance* (9):565–573.

183. S. D. Caruthers, A. M. Neubauer, et al. 2006. In vitro demonstration using ^{19}F magnetic resonance to augment molecular imaging with paramagnetic perfluorocarbon nanoparticles at 1.5 tesla. *Investigative Radiology* (41):305–312.

184. M. Modo et al. 2004. Mapping transplanted stem cell migration after a stroke: A serial, in vivo magnetic resonance imaging study. *Neuroimage* (21):311–317.

185. R. E. Jacobs, S. E. Fraser. 1994. Magnetic-resonance microscopy of embryonic-cell lineages and movements. *Science* (263):681–684.

186. T. C. Yeh, W. Zhang, S. T. Ildstad, C. Ho. 1993. Intracellular labeling of T-cells with superparamagnetic contrast agents. *Magnetic Resonance Medicine* (30):617–625.

187. E. T. Ahrens, M. Feili-Hariri, H. Y. Xu, G. Genove, P. A. Morel. 2003. Receptor-mediated endocytosis of iron-oxide particles provides efficient labeling of dendritic cells for in vivo MR imaging. *Magnetic Resonance Medicine* (49):1006–1013.

188. M.F. Kircher et al. 2003. In vivo high resolution three-dimensional imaging of antigen-specific cytotoxic T-lymphocyte trafficking to tumors. *Cancer Research* (63):6838–6846.

189. A. Stroh, C. Faber, T. Neuberger, P. Lorenz, K. Sieland, P. M. Jakob, A. Webb, et al. 2005. In vivo detection limits of magnetically labeled embryonic stem cells in the rat brain using high-field (17.6 T) magnetic resonance imaging. *Neuroimage* (24):635–645.

190. J. W. M. Bulte, A. S. Arbab, T. Douglas, J. A. Frank. 2004. Preparation of magnetically labeled cells for cell tracking by magnetic resonance imaging. *Methods Enzymology* (386):275–299.

191. D. L. Kraitchman, A. W. Heldman, E. Atalar, L. C. Amado, B. J. Martin, M. F. Pittenger, J. M. Hare, et al. 2003. In vivo magnetic resonance imaging of mesenchymal stem cells in myocardial infarction. *Circulation* (107):2290–2293.

192. J. M. Hill, A. J. Dick, V. K. Raman, R. B. Thompson, Z. X. Yu, K. A. Hinds, B. S. Pessanha, et al. 2003. Serial cardiac magnetic resonance imaging of injected mesenchymal stem cells. *Circulation* (108):1009–1014.

193. J. W. Bulte, D. L. Kraitchman. 2004. Monitoring cell therapy using iron oxide MR contrast agents. *Current Pharmaceutical Biotechnology* (5):567–584.

194. C. H. Cunningham, T. Arai, P. C. Yang, M. V. McConnell, J. M. Pauly, S. M. Conolly. 2005. Positive contrast magnetic resonance imaging of cells labeled with magnetic nanoparticles. *Magnetic Resonance Medicine* (53):999–1005.

195. E. T. Ahrens, R. Flores, H. Xu, P. A. Morel. 2005. In vivo imaging platform for tracking immunotherapeutic cells. *Nature Biotechnology* (23):983–987.

196. K. C. Partlow, J. Chen, et al. 2007. ^{19}F magnetic resonance imaging for stem/progenitor cell tracking with multiple unique perfluorocarbon nanobeacons. *FASEB Journal* (21):1647–1654.

197. A. S. Arbab, G. T. Yocum, L. B. Wilson, A. Parwana, E. K. Jordan, H. Kalish, J. A. Frank. 2004. Comparison of transfection agents in forming complexes with ferumoxides, cell labeling efficiency, and cellular viability. *Molecular Imaging* (3):24–32.
198. (a) S. Zhang, M. Merritt, D. E. Woessner, R. E. Lenkinski, A. D. Sherry. 2003. *Accounts of Chemical Research* (36):783; (b) K. M. Ward, A. H. Aletras, R. S. Balaban. 2000. *Journal of Magnetic Resonance Imaging* 143:79.
199. J. Stancanello, E. Terreno, D. D. Castelli, C. Cabella, F. Uggeri, S. Aime. 2008. Development and validation of a smoothing-splines-based correction method for improving the analysis of CEST-MR images. Contrast media. *Molecular Imaging* Aug 6 (online publication).
200. J. M. Zhao, Y. E. Har-el, M. T. McMahon, J. Zhou, A. D. Sherry, G. Sgouros, J. W. Bulte, et al. Size-induced enhancement of chemical exchange saturation transfer (CEST) contrast in liposomes. *Journal of American Chemical Society* 130 (15): 5178–5184.
201. E. Terreno, D. D. Castelli, L. Milone, S. Rollet, J. Stancanello, E. Violante, S. Aime. 2008. First ex-vivo MRI colocalization of two LIPOCEST agents. *Contrast Media Molecular Imaging* 3 (1): 38–43.
202. E. Terreno, A. Barge, L. Beltrami, G. Cravotto, D. D. Castelli, F. Fedeli, B. Jebasingh, et al. 2008. Highly shifted LIPOCEST agents based on the encapsulation of neutral polynuclear paramagnetic shift reagents. *Chemical Communications (Cambridge)* (5):600–602.
203. E. Terreno, C. Cabella, C. Carrera, D. Delli Castelli, R. Mazzon, S. Rollet, J. Stancanello, et al. 2007. From spherical to osmotically shrunken paramagnetic liposomes: An improved generation of LIPOCEST MRI agents with highly shifted water protons. *Angewandt Chemie* International Edition England 46 (6): 966–968.
204. C. Adair, M. Woods, et al. 2007. Spectral properties of a bifunctional PARACEST europium chelate: An intermediate for targeted imaging applications. *Contrast Media Molecular Imaging* (2):55–58.
205. P. M. Winter, K. Cai, et al. 2006. Targeted PARACEST nanoparticle contrast agent for the detection of fibrin. *Magnetic Resonance in Medicine* (56):1384–1388.
206. D. Pan, X. Cai, B. Kim, A. J. Stacy, L. V. Wang, G. M. Lanza. 2012. Rapid synthesis of near infrared polymeric micelles for real-time sentinel lymph node imaging. *Advanced Healthcare Materials* 1:582–589.
207. D. Pan, X. Cai, C. Yalaz, A. Senpan, K. Omanakuttan, S. A. Wickline, L. V. Wang, et al. 2012. Photoacoustic sentinel lymph node imaging with self-assembled copper neodecanoate nanoparticles. *ACS Nano* 6:1260–1267.
208. X. Wang, Y. Pang, G. Ku, X. Xie, G. Stoica, L. V. Wang. 2003. Noninvasive laser-induced photoacoustic tomography for structural and functional in vivo imaging of the brain. *Nature Biotechnology* 21:803–806.
209. M. L. Li, J. T. Oh, X. Y. Xie, G. Ku, W. Wang, C. Li, G. Lungu, et al. 2008. Simultaneous molecular and hypoxia imaging of brain tumors in vivo using spectroscopic photoacoustic tomography. *Proceedings of the IEEE* 96:481–489.
210. X. D. Wang, X. Y. Xie, G. N. Ku, L. H. V. Wang. 2006. Noninvasive imaging of hemoglobin concentration and oxygenation in the rat brain using high-resolution photoacoustic tomography. *Journal of Biomedical Optics* 11.
211. G. F. Lungu, M. L. Li, X. Y. Xie, L. H. V. Wang, G. Stoica. 2007. In vivo imaging and characterization of hypoxia-induced neovascularization and tumor invasion. *International Journal of Oncology* 30:45–54.
212. G. Ku, X. D. Wang, G. Stoica, L. H. V. Wang. 2004. Multiple-bandwidth photoacoustic tomography. *Physics in Medicine and Biology* 49:1329–1338.
213. A. Agarwal, S. W. Huang, M. O'Donnell, K. C. Day, M. Day, N. Kotov, S. Ashkenazi. 2007. Targeted gold nanorod contrast agent for prostate cancer detection by photoacoustic imaging. *Journal of Applied Physics* 102.

214. A. De la Zerda, C. Zavaleta, S. Keren, S. Vaithilingam, S. Bodapati, Z. Liu, et al. 2008. Carbon nanotubes as photoacoustic molecular imaging agents in living mice. *Nature Nanotechnology* 3:557–562.

215. W. Li, P. K. Brown, L. V. Wang, Y. Xia. 2011. Gold nanocages as contrast agents for photoacoustic imaging. *Contrast Media Molecular Imaging* 6:370–377.

216. G. Ku, M. Zhou, S. Song, Q. Huang, J. Hazle, C. Li. 2012. Copper sulfide nanoparticles as a new class of photoacoustic contrast agent for deep tissue imaging at 1064 nm. *ACS Nano* 6:7489–7496.

217. V. Ntziachristos, C. Bremer, R. Weissleder. 2003. Fluorescence imaging with near-infrared light: New technological advances that enable in vivo molecular imaging. *European Radiology* 13:195–208.

218. R. Weissleder, V. Ntziachristos. 2003. Shedding light onto live molecular targets. *Nature Medicine* 9:123–128.

219. E. I. Altinoglu, J. H. Adair. 2010. Near infrared imaging with nanoparticles. *Wiley Interdisciplinary Reviews of Nanomedicine and Nanobiotechnology* 2:461–477.

220. D. Pan, M. Pramanik, A. Senpan, J. S. Allen, H. Zhang, S. A. Wickline, L. V. Wang, et al. 2011. Molecular photoacoustic imaging of angiogenesis with integrin-targeted gold nanobeacons. *FASEB Journal* 25:875–882.

2 Nuclear Imaging with Nanoparticles

Ali Azhdarinia and Sukhen C. Ghosh

CONTENTS

2.1 INTRODUCTION

Nuclear medicine is a medical specialty that uses radioactive substances for diagnosis and therapy. It is recognized as a powerful noninvasive modality capable of acquiring whole-body images and revealing changes in functional activity within tissues. Clinically, nuclear imaging procedures are used for diagnosis of many types of cancer, heart disease, neurological disorders, and other abnormalities [1–3]. The major advantages of nuclear imaging over other molecular imaging modalities are its exquisite detection sensitivity (femto- to picomolar range) and the ability to image without concerns over tissue penetration. Biomedical radioisotopes, used either alone or as part of a radiolabeled molecule, emit photons with energies ranging from 30 to 511 keV, which are detected by specialized camera systems. Based on the emission properties of the radioisotope and its detection method, nuclear imaging is classified as either positron emission tomography (PET) or single-photon emission computed tomography (SPECT). Both techniques involve the use of functional contrast agents to generate three-dimensional tomographic images used for diagnosis, treatment planning, and monitoring therapeutic response.

The growing interest in molecular imaging with PET and SPECT is attributable to several factors. First, the development of hybrid scanners has enabled the functional imaging capabilities of PET and SPECT to be combined with anatomical correlation obtained through computed tomography (CT), and, more recently, with magnetic resonance imaging (MRI). Fused images (i.e., PET/CT or SPECT/CT) have improved diagnostic accuracy and have affected clinical management in a significant proportion of patients with a wide range of diseases by guiding subsequent procedures, excluding the need for unnecessary procedures, changing both inter- and intramodality therapy, and providing prognostic information [4]. The clinical impact of hybrid molecular/anatomical imaging has extended into preclinical research in the form of highly sensitive small-animal imaging systems that are capable of resolving structures below 1 mm [5]. Second, the discovery of new molecular targets has given rise to cross-disciplinary programs involving molecular biology, genetics, chemistry, bioengineering, and clinical medicine to develop target-specific molecules for disease detection. Such diversity in research and clinical expertise allows for (1) designation of high-value clinical targets, (2) innovative drug design strategies, (3) agent validation at the molecular level, and (4) testing in clinically useful models of disease. Third, the increased focus on translational research has encouraged basic science laboratories to integrate molecular imaging procedures into their characterization process for new molecular entities and has given rise to a growing toolkit of molecular imaging agents.

Alongside the continued growth of conventional molecular imaging agents, nanoparticles have emerged as an attractive drug delivery strategy and are becoming increasingly important in medicine. By labeling with radioactive contrast, a nanoparticle drug delivery platform can be monitored in animal models to assess its biodistribution, pharmacokinetic (PK) profile, and overall effectiveness for therapy. Well-defined labeling chemistries for conventional biomolecules (i.e., small molecules, peptides, and antibodies) and imaging protocols that have been clinically validated have been adapted for nanotechnologies to enable the rational development and characterization of radioactive nanoparticles. In this chapter, we describe the importance of nuclear imaging in the development and characterization of nanoparticle-based imaging agents. We review various classes of radiolabeled nanoparticles and discuss key components involved in agent design and optimization.

2.2 PRINCIPLES OF NUCLEAR IMAGING

2.2.1 PET

PET is a highly sensitive, three-dimensional tomographic imaging technique initially used as a research tool in humans in the 1970s, and has become a vital component of patient care in several disease types [6]. Its unique tomographic capability comes from the simultaneous emission of two 511 keV gamma rays that travel in opposite directions approximately 180° apart as a result of the interaction of a positron with an electron in an annihilation event. PET coincidence detection registers the annihilated photons to reconstruct their source along the line of collection and provides clinicians with the ability to better determine and quantify regional function and metabolism in most anatomic sites. For example, in the brain, PET is used to determine grade, type, and proliferative activity of gliomas [7], provides early detection, differential diagnosis, prognosis, and follow-up of dementia [8,9], and is important for imaging of movement disorders such as Parkinson's disease, multiple system atrophy (MSA), and progressive supranuclear palsy (PSP) [10]. In cardiovascular disease, PET enables diagnosis and management of patients with known or suspected coronary artery disease through myocardial perfusion imaging with FDA approved rubidium-82 (^{82}Rb) [11] or nitrogen-13 (^{13}N) labeled ammonia [12], as well as assessment of myocardial viability with the metabolic imaging agent fluorine-18 (^{18}F) labeled deoxyglucose (^{18}F-FDG) [13]. In cancer, ^{18}F-FDG-PET is considered the gold standard for tumor imaging and is used to visualize the metabolic differences between normal and cancer cells [14,15]. ^{18}F-FDG also possesses well-defined guidelines for production and quality assessment that serve as guides for new radiotracers poised to enter clinical studies [16]. Numerous other targeting agents employ PET imaging to obtain critical information about proliferation, vasculature, receptor status, hypoxia, and apoptosis [17–20]. Furthermore, the importance of characterizing drug delivery platforms, such as nanoparticles, for in vivo diagnosis and therapy suggests PET will play a vital role in agent validation and will guide developmental strategies. Given its versatility to perform clinical assessment of small molecule metabolic imaging, receptor imaging with gallium-68 (^{68}Ga) [21] and copper-64 (^{64}Cu) labeled peptides [22] (Figure 2.1), and immunoPET with radiolabeled antibodies [23], PET is expected to play a prominent role in the development and validation of radiolabeled nanoparticles for personalized medicine.

The high spatial resolution of PET (1–2 mm in humans) is dependent upon several factors, including positron range. Thus, radionuclide selection must be carefully considered to optimize image quality [24]. PET has benefited from the availability of a robust group of positron-emitting radionuclides that have distinct decay properties,

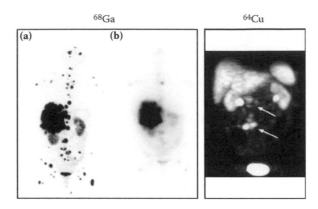

FIGURE 2.1 Left: (a) PET imaging with ⁶⁸Ga-DOTA-TOC in a 56-y-old woman with multiple liver and lymph node metastases clearly depicted visceral metastases, and osteoblastic and osteolytic bone metastases. (b) Only some of these bone metastases were delineated by conventional scintigraphy. Right: PET imaging with ⁶⁴Cu-TETA-OC identified abdominal carcinoid lesions (arrows) in a patient with carcinoid tumors. (Adapted with permission from Gabriel, M. et al. 2007. *Journal of Nuclear Medicine* 4:508–518 and Anderson, C. J. et al. 2001. *Journal of Nuclear Medicine* 2:213–221.)

half-lives, and labeling chemistries. This affords flexibility in radiotracer design and enables the use of different classes of compounds for targeting, such as analogs of biological molecules, synthetic compounds, and nanoparticles. Due to differences in targeting properties (active vs. passive), solubility, and biocompatibility, nanoparticle-based agents can require vastly different imaging time windows (i.e., from 6 hr up to several days postinjection) for adequate evaluation of their in vivo properties. Therefore, delayed imaging (>24 hr) is generally preferred to allow for maximal clearance from nontarget sites in order to obtain optimal contrast and provide a more representative evaluation of agent fate in biological systems. While this prohibits the use of common PET radionuclides such as carbon-11 (^{11}C) and ^{13}N, it aligns well with radionuclides applied in antibody imaging such as ^{64}Cu, iodine-124 (^{124}I), zirconium-89 (^{89}Zr), and yttrium-86 (^{86}Y). It should be noted that these radionuclides are produced by a cyclotron facility and may not be locally available to many end users, but their sufficiently long half-lives make shipping for next-day use practical.

2.2.2 SPECT

Single-photon radiotracers account for the majority of nuclear medicine procedures performed owing to the large installation base of gamma cameras in clinics worldwide. Images can be acquired with conventional two-dimensional (2D) gamma imaging (referred to as planar imaging or scintigraphy) using gamma cameras invented in the middle of the last century by Anger [25], or by three-dimensional (3D) tomographic imaging. Unlike PET, SPECT has traditionally been considered as a nonquantitative modality. However, advances in multimodality gamma cameras (SPECT/CT), algorithms for image reconstruction, and sophisticated compensation techniques to correct for photon attenuation and scattering have now made quantitative SPECT viable in a manner similar to quantitative PET (i.e., kBq·cm^{-3}, standardized uptake value) [26].

In SPECT, a collimation system is placed in front of the detectors to determine the origin of a photon. This is an essential component of image acquisition, but it results in the exclusion of a large number of emitted photons and significantly reduces the sensitivity of SPECT compared to PET, which does not use collimation [27]. However, SPECT does possess several advantages that make it an important modality in molecular imaging and drug discovery. For example, a long list of clinical SPECT imaging procedures are routinely performed, including myocardial perfusion imaging, functional brain imaging, renal scans, bone imaging, thyroid scans, lymphoscintigraphy, and cancer imaging [28] (Figure 2.2). From these, radiolabeling techniques have been developed, optimized to provide the highest yields, and often placed in kit form to simplify preparation. As a result, clinically validated labeling approaches have been made available for use with new molecular imaging agents, as well as in the development of nanoparticle imaging platforms, for in vivo characterization. Another advantage of SPECT is the availability of isotopes that do not rely on cyclotron production. Among these is technetium-99m (^{99m}Tc, $t_{1/2}$ = 6.01 hr), a generator-produced radiometal that is the most widely used radionuclide in nuclear medicine because of its favorable imaging properties, low cost, and availability within standard radiopharmacies. Chelating agents used for ^{99m}Tc labeling are well known and have a long history of clinical use, making them a proven commodity

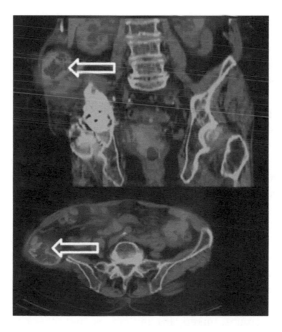

FIGURE 2.2 (See color insert.) SPECT/CT imaging with ^{131}I-iodide in a patient with differentiated follicular thyroid carcinoma after pelvic surgery shows a benign tracer accumulation in the colon (arrow). (Adapted with permission from Mariani, G. et al. 2010. *European Journal of Nuclear Medicine and Molecular Imaging* 10:1959–1985.)

within the development strategy of new imaging agents. Indium-111 (^{111}In) is another clinically relevant SPECT radionuclide and has a long half-life (2.8 d) that is suitable for studying molecules with long circulation times, such as nanoparticles, and also possesses well-developed radiochemistry schemes for new agent development. SPECT also possesses the unique capability to conduct simultaneous multitracer studies by using radionuclides with different energies and half-lives. This permits investigation of several biological processes within a single imaging session, a task that is not possible with PET. With the growing importance of preclinical imaging in drug discovery, the established SPECT infrastructure is expected to continue its vital involvement in agent design, characterization, and translation.

2.2.3 RADIOTRACERS

Nuclear imaging is predicated on the use of radioactive contrast agents, or radiotracers, that are administered in subpharmacologic doses and follow the tracer principle, a fundamental component of nuclear imaging discovered by George de Hevesy in 1913. The tracer principle allows monitoring of the in vivo distribution of a compound without disturbing body function. A radiotracer can be as simple as a radioactive element (i.e., ^{18}F for bone scans) or a radioactive molecule (i.e., oxygen-15 labeled water for measurement of myocardial blood flow) that lacks molecular specificity, but is capable of providing valuable diagnostic information. Molecularly targeted radiotracers, on the other hand, are more intricate in design and consist of a targeting moiety, a radiolabel, and various linkers or functional groups that permit radiolabeling without disrupting biological activity. Targeting moieties are generally categorized as small molecules, peptides, proteins, or antibodies, and they can be used for detecting components of metabolic and biochemical pathways, cell death, angiogenesis, reporter gene expression, cell proliferation, and receptor expression levels.

For a given type of targeting compound, unique chemical and biological properties must be considered during agent design to maximize effectiveness. Antibodies are large molecules (150 kDa) with exceptional affinity for cell surface receptors and extracellular antigens and are widely used in molecular imaging. Due to their large size, antibodies can undergo conjugation to linker groups or radiometal chelating agents without significant loss in binding affinity, but reactions must proceed near physiologic pH and temperature to avoid degradation, thus limiting the types of radiolabeling strategies that can be used. Conversely, peptides can undergo harsher radiolabeling reactions that occur at low pH (i.e., pH 3–4) and require heating at 90°C–100°C, but require careful selection of conjugation sites for linkers or chelating agents to avoid steric effects that may reduce binding potency. Small molecule imaging agents have the advantage of production via organic synthesis, which adds tremendous flexibility in design by allowing chemists to select from an immense pool of protecting groups, linkers, leaving groups, and other chemical compounds that can be used for agent customization. As with peptides, however, the addition of a radioactive label to the small molecule must be carried out in a manner that does not impact binding and preserves favorable PK characteristics.

2.3 THERANOSTIC ROLE OF NUCLEAR IMAGING

In addition to diagnosis, nuclear medicine also acts as a theranostic modality (i.e., uses diagnostic properties to personalize therapy). For example, the radiolabeled analog of a therapeutic compound can aid in selecting patients that are likely to respond to a particular treatment by confirming the presence of the molecular target prior to therapy. This has been shown by PET imaging with 16α-[^{18}F]-fluoro-17β-estradiol (^{18}F-FES), a radiolabeled estradiol analog that detects and monitors estrogen receptor (ER) status in patients with advanced ER+ breast cancer, and has demonstrated the ability to predict responsiveness to tamoxifen therapy [29]. In neuroendocrine tumors, PET imaging with ^{68}Ga-labeled somatostatin (SST) analogs successfully identified patients who were likely to benefit from peptide receptor radionuclide therapy using beta emitters labeled with the same SST analog, and also monitored potential changes in somatostatin receptor (SSTR) abundance over the course of the therapy [30]. In an alternative theranostic approach, therapeutic payloads can be targeted with different vectors and radiolabeled to facilitate characterization of in vivo properties and efficacy. Later, we discuss the advantages and disadvantages of this strategy with different types of targeting biomolecules.

2.3.1 DRUG DELIVERY WITH CONVENTIONAL BIOMOLECULES

Despite tremendous advances in molecular targeting approaches for cancer therapy, several malignancies are still treated with chemotherapy agents that are highly cytotoxic and lack tumor specificity. As a result, severe side effects arise that are dose limiting and decrease the effective therapeutic window. Strategies to combine antineoplastic agents with a targeting moiety have long been sought to produce bioconjugates that actively home to a specific molecular target and reduce off-target effects. One approach for improving tumor localization of cytotoxic agents is to link them chemically to antibodies that possess high affinity for a tumor-associated antigen or an antigen that may reside in the peritumoral space. Advances in antibody engineering offer great versatility in the design of these targeting vectors and give improved control over size (and thus PK properties) (Figure 2.3) as well as insertion of reactive moieties that permit site-specific conjugation [31–33]. The development of antibody–drug conjugates (ADCs) has expanded the benefits of antibody targeting and is recognized as an attractive therapeutic approach owing to the ability of antibodies to tolerate the conjugation of low molecular weight (MW) chemotherapy agents without losing bioactivity [34]. ADCs can be generated by attaching a cytotoxic compound to lysine residues that are randomly distributed throughout the antibody, or through a site-specific approach that binds to reduced interchain disulfides and ensures that conjugation does not impact the binding region [35]. However, drawbacks such as immunogenicity and ability to diffuse through biological barriers are concerns associated with antibody-based targeting. In addition, the potential for antibodies to denature under high heat, high or low pH conditions, and exposure to certain organic solvents greatly limits the scope of chemical modification they can undergo and impacts the broad application of their use for drug delivery.

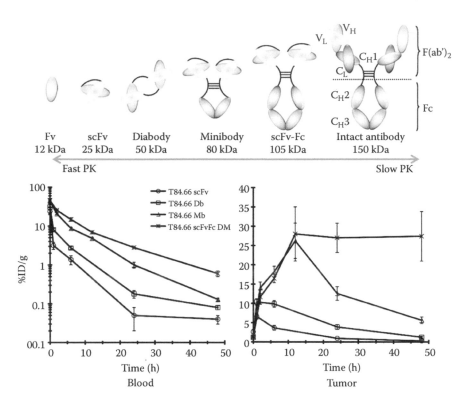

FIGURE 2.3 Top: Schematic presentation of an intact antibody and engineered antibody fragments derived from it including a single variable domain fragment (Fv), single chain Fv (scFv), diabody, minibody, and scFv-Fc. Molecular weights are indicated below each fragment. V_L = variable light; V_H = variable heavy, C_L = constant light; C_H = constant heavy. Bottom: Blood clearance (left) and tumor uptake (right) curves of radioiodine-labeled anti-CEA T84.66 antibody fragments in tumor-bearing mice. The longer persistence of radiolabeled fragments in the circulation leads to higher and more persistent tumor uptakes, but longer times are required to obtain high-contrast imaging. (Adapted with permission from Olafsen, T. and A. M. Wu. 2010. *Seminars in Nuclear Medicine* 3:167–181.)

Another method to improve the delivery of chemotherapy is through the use of targeted peptides. Peptides already play an active and diverse role in the clinical management of cancer, evidenced by their application in both diagnosis and therapy [36–40], and they are attractive for bioconjugate development for several reasons. First, the discovery of novel peptides by phage display provides a robust pipeline of targeting vectors with potential applications in imaging. Additionally, peptides have high affinity for cell surface receptors and are rapidly cleared from background tissues, thus producing images with increased contrast. Furthermore, peptide manufacturing by solid-phase peptide synthesis (SPPS) is well controlled, customizable for small or bulk quantities, allows for site-specific conjugation, and enables efficient purification and characterization methods. Despite the advantages of developing peptide–drug conjugates, several challenges exist. For example, peptides are structurally complex molecules as they have numerous functional groups that are susceptible to enzymatic degradation and inactivation. They also have a short plasma half-life, which may not be ideal for delivering a desired dose to the target and may require tuning through the use of a PK modifier within the final bioconjugate. There is also concern over possible steric effects that may impair the binding properties of a small peptide (e.g., MW 1500 Da) after conjugation with a chemotherapy agent such as paclitaxel (MW 854 Da). This is further complicated if a multimeric construct containing several chemotherapy moieties is desired and could also contribute to reduced solubility. While solubility issues are generally addressed by incorporating hydrophilic linkers such as polyethylene glycol (PEG), the complexity of the agent design becomes greater and the inherent advantages of favorable PK and simple manufacturing may be negated.

Targeting of nonspecific chemotherapy agents can also be achieved using small molecule drug conjugates. Generally, the small molecule used for targeting is biochemically active, is nonimmunogenic, and can navigate through biological barriers to achieve target localization and enhance therapeutic effects. In addition, small molecules can be readily modified by organic synthesis routes, enabling tremendous versatility in drug design. The most widely used approach for small molecule drug targeting uses folate (also known as vitamin B9) to exploit the expression of folate receptors on the plasma membrane of cancer cells. Liu et al. described a novel conjugate comprising folate and the chemotherapy agent, 5-fluorouracil (5-FU), which proved to be superior to a nontargeted 5-FU analog and monomeric 5-FU in colorectal and 5-FU-resistant colorectal cells [41]. However, as is the case with peptides, small molecules are likely to incur significant changes in binding properties, solubility, biodistribution, and PK following conjugation with a therapeutic compound and must be carefully designed to maximize their effectiveness for drug delivery.

2.3.2 DRUG DELIVERY WITH NANOPARTICLES

Nanotechnology is defined as the intentional design, characterization, production, and application of materials, structures, devices, and systems by controlling their size and shape in the nanoscale range (1 to 100 nm) [42]. To address the significant limitations of conventional bioconjugate approaches described earlier, nanoparticle systems have been implemented as drug delivery vehicles to maximize the efficacious properties of targeting, diagnostic, and cytotoxic compounds. Because of their nanometer size, nanoparticles are adept at overcoming biological barriers that often limit the effectiveness of other types of molecules. Advances in surface chemistry with polymeric systems such as PEG and other hydrophilic moieties have allowed these vectors to become more biocompatible and overcome high accumulation in organs of the reticuloendothelial system (RES). Furthermore, tuning of PK has been achieved through surface chemistry to optimize in vivo properties. The amenability of nanoparticles to chemical modification is attributed to their high surface area-to-volume ratios, which permit high levels of surface functionalization. This characteristic affords the ability to develop multivalent constructs to increase sensitivity

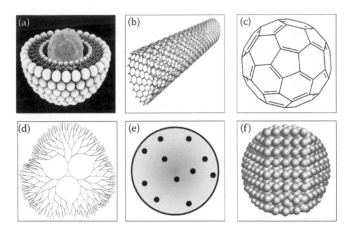

FIGURE 2.4 Selection of nanomaterials commonly used in medicine. (a) liposome, (b) single-walled carbon nanotube, (c) fullerene, (d) dendrimer, (e) iron oxide nanoparticle, and (f) quantum dot.

for a particular target and to perform functionalization with different constituents (e.g., contrast agents) to produce multimodality agents. Examples of nanoparticles commonly used in medicine are shown in Figure 2.4.

Nuclear imaging techniques play an important role in the advancement of nanoparticles in several ways. First, the ability to load (or internalize) radionuclides permits in vivo imaging of nanoparticles and noninvasive assessment of their utility for drug delivery. Second, the radiolabeling efficiency of nanoparticles has improved with chelation schemes adapted from clinical nuclear medicine and has resulted in simpler agent formulation with images that more accurately reflect biodistribution, PK, and targeting ability. Third, the application of surface modification methods has created new routes to improve solubility and biocompatibility and has directly impacted imaging characteristics. Fourth, targeting moieties introduced by optimized surface chemistry have enabled active targeting of nanoparticles to diseases such as cancer. By integrating (1) experiences with clinically relevant radionuclides, (2) advances in radiopharmaceutical chemistry, and (3) emerging targeting strategies for molecular imaging into their development, radiolabeled nanoparticles are viewed as a promising class of agents for diagnosis and therapy.

2.4 RADIONUCLIDE SELECTION

Several biomedical radionuclides with unique decay properties are suitable for use with nanoparticles and are listed in Table 2.1. Given that nanoparticles can accommodate different types of radiolabeling schemes, a systematic approach for radionuclide selection is necessary in order to maximize the imaging utility of a radiolabeled nanoparticle. The criteria for radionuclide selection begin with identification of the desired nuclear imaging modality. In recent years, device companies have developed preclinical scanners with hybrid capabilities enabling PET/SPECT/CT imaging from a single instrument. Access to these systems gives radiochemists maximum

TABLE 2.1

Characteristics of PET and SPECT Radionuclides Used for Radiolabeling Nanoparticles

Radionuclide	Emission Type	Half-Life	E_{max} γ (keV)
^{64}Cu	β^+	12.7 hr	657
^{18}F	β^+	109.8 min	640
^{68}Ga	β^+	67.6 min	1899
^{124}I	β^+	4.2 d	2130
^{89}Zr	β^+	3.3 d	897
^{67}Ga	γ, Auger	3.3 d	393
^{123}I	γ, Auger	13.3 hr	159
^{125}I	γ	60.1 d	35
^{131}I	γ, β^-	8.0 d	364
^{111}In	γ, Auger	2.8 d	245
^{99m}Tc	γ	6.01 hr	140

flexibility in designing radiolabeled nanoparticles and allows the use of any available diagnostic radionuclide. Conversely, should a laboratory be limited to a single type of nuclear scanner, the choice of radionuclide is restricted accordingly.

The half-life of a radionuclide plays an important role in radiosynthesis, imaging applications, and dosimetry. For example, ^{11}C is a short-lived ($t_{1/2}$ = 20.4 min) positron emitter that is relegated to on-site use and must be imaged shortly after injection to a subject. Other PET radionuclides, such as ^{18}F ($t_{1/2}$ = 109.8 min) and ^{68}Ga ($t_{1/2}$ = 67.6 min), possess longer half-lives that allow for extended radiochemical procedures and quality analysis, as well as shipping of doses to sites within reasonable range of the production facility. With typical imaging time points at 1 hr postinjection, ^{18}F and ^{68}Ga are not effective for delayed imaging and thus have limited utility with imaging agents that are slowly cleared from circulation, including certain types of nanoparticles. Alternatively, radionuclides such as ^{99m}Tc ($t_{1/2}$ = 6.01 hr), ^{111}In ($t_{1/2}$ = 2.8 days), and ^{64}Cu ($t_{1/2}$ = 12.7 hr) possess half-lives better suited for delayed imaging and are therefore considered to be more useful for studying nanoparticles. Since the PK profile of a nanoparticle can be significantly altered by coating or surface modifications, comparison of multiple radiolabeled versions of a particular agent may be warranted in order to select the optimal radionuclide strategy for in vivo imaging. It is also important to consider the effect of half-life and energy of the emitted particles on the radiation dose delivered to the subject to ensure that each imaging procedure does not generate excessive radiation doses.

Radionuclides can be attached to biomolecules either directly or through the presence of a chelating agent. Direct attachment is generally associated with radiohalogens (i.e., ^{18}F or ^{124}I) that are covalently bound to the side chain of amino acid residues, or to reactive groups that are chemically introduced into the molecule. Radiometals, on the other hand, require the use of chelating agents (Figure 2.5) to form coordination complexes using various combinations of electron-donating

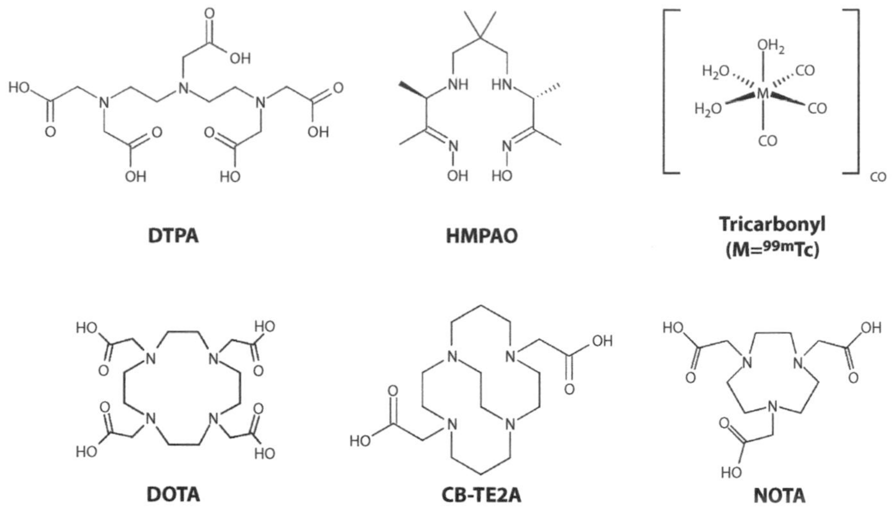

FIGURE 2.5 Examples of chelating agents used for radiolabeling.

atoms such as nitrogen and oxygen. When using radiometals, one must also consider that each chelating agent has its own unique radiolabeling procedures. For example, ^{64}Cu can be efficiently chelated with mild heat or may require temperatures to be as high as 95°C depending on chelator selection [43]. Similar diversity is observed in the pH range of reaction buffers, which can be acidic, neutral, or basic for different chelators. Thus, the chelation approach must also take into account the compatibility of the nanoparticle system with the radiolabeling conditions. Strategies for radio-labeling nanoparticles have followed the successes of conventional imaging agents and will likely become even more robust as new molecular imaging agents undergo clinical evaluation and demonstrate efficacy in patients.

Identification of the ideal radiolabeling approach must also take into account the availability of the radionuclide. For example, radionuclides such as ^{64}Cu, ^{89}Zr, and ^{124}I are attractive candidates for use with nanoparticles owing to their long half-lives and PET capabilities; however, they can only be produced by a cyclotron facility. This drawback is not completely restrictive however, as their half-lives allow for shipping to off-site laboratories, though limitations in production scale and schedule must be taken into account during coordination of studies. Alternatively, radionuclides can be produced by generator systems (i.e., ^{99m}Tc, ^{68}Ga) housed directly in a labora-tory or radiopharmacy and may be eluted multiple times per day to collect radio-activity. Compared to cyclotron-produced radionuclides, this "on-demand" access to radioactivity promotes faster method development as radiochemistry techniques can be iteratively tested and optimized in a rapid fashion. The abundance of studies that use ^{99m}Tc for labeling nanoparticles attests to the value of generator-produced radionuclides. Furthermore, planning of imaging studies with generator-based radio-nuclides becomes much more flexible as they are not dependent on defined pro-duction schedules that may exist for cyclotrons. Finally, radionuclide selection can take advantage of existing radioisotope infrastructure, such as that of ^{111}In, which is readily available through commercial sources, uses kit-based chemistry, and has been extensively used with different types of imaging agents [44,45]. Later, we describe radionuclides that are useful for nanoparticle imaging and focus on concepts related to emission properties and labeling strategies for in vivo applications.

2.4.1 TECHNETIUM-99M

^{99m}Tc is a single-photon emitter that is considered the workhorse of nuclear medicine on the basis of its use in more than 80% of the 23–25 million diagnostic doses received by patients annually in the United States [46]. ^{99m}Tc-labeled radiotracers have diverse clinical applications including assessment of renal function, bone scans, brain perfusion, myocardial imaging, and detection of infection and inflammation [47]. Among the many attractive characteristics of ^{99m}Tc, the most significant factor in its widespread use is its production by the ^{99}molybdenum/^{99m}technetium (^{99}Mo/^{99m}Tc) generator that was first discovered in 1958 by Tucker and Greene. With the continuously growing demand for diagnostic procedures in the clinic, the ^{99}Mo/^{99m}Tc generator plays a vital role in nuclear medicine as it is available worldwide, is inexpensive, and can be housed within a standard radiopharmacy. Another key attribute of ^{99m}Tc is its favorable decay properties, including its low-energy gamma emission, which is convenient for detection and does not impart excessive radiation doses to patients. Also, its 6.01 hr half-life is long enough to allow for synthesis of a radiopharmaceutical, assessment of quality control, and imaging at delayed time points, making it compatible with different classes of targeting agents. The ready availability of ^{99m}Tc has been coupled with "instant kits" to simplify labeling practices and permit rapid preparation of clinical doses.

Radiolabeling protocols for ^{99m}Tc agents generally rely on the use of coordination complexes with chelating agents containing varying combinations of electron donating atoms such as nitrogen, oxygen, sulfur, and phosphorus. ^{99m}Tc complexes are highly stable in vivo and useful for labeling small molecules, peptides, proteins, antibodies and their fragments, and several nanoparticle platforms that will be discussed later in this chapter.

2.4.2 INDIUM-111

^{111}In is a single-photon emitting radiometal that has a long history of use in nuclear medicine. Its 2.8 d half-life is ideal for in vivo monitoring of compounds that are slowly cleared from circulation and require imaging up to several days after injection in order to obtain adequate contrast. In the clinic, ^{111}In-labeled antibodies are used in patients with prostate cancer [48] and lymphoma [49]; thus, standardized synthesis and imaging protocols are available and can be applied to novel imaging compounds. In addition to its gamma rays used for diagnosis, ^{111}In emits Auger electrons, which are useful for specific tumor cell killing with a low level of damage to surrounding cells [50], thus giving it theranostic capabilities. The combination of diagnostic and radiotherapeutic potency makes ^{111}In an attractive radionuclide for use with nanoparticles since it can be used to track in vivo properties at low (diagnostic) doses and can also be prepared at significantly higher therapeutic amounts and combined with cytotoxic payloads to provide combination therapy. It is significant to note that ^{111}In possesses proven chelation techniques with diethylenetriaminepentaacetic acid (DTPA) and 1,4,7,10-tetraazacyclododecane-1,4,7,10-tetraacetic acid (DOTA) that allow efficient radiolabeling with excellent in vivo stability and obviates the need for extensive method development during application to new compounds. These characteristics, coupled with its unique decay properties, have made ^{111}In an important radionuclide in the development of nanoparticle-based imaging agents.

2.4.3 COPPER-64

[64]Cu is a cyclotron-produced positron emitter that is widely used in antibody labeling, but has also grown in importance for labeling of nanoparticles. It has a 12.7 hr half-life that is suitable for delayed imaging up to 48 hr after injection and has well-defined chelation chemistry with tetraaza chelating agents. DOTA is most frequently used for ^{64}Cu-radiolabeling owing to its low cost, mild conditions (pH 5.5–6, 40°C) needed for metal complexation, and ability to chelate several radiometals should comparative studies on the same conjugate be desired. However, the in vivo stability of ^{64}Cu-DOTA is somewhat limited due to transchelation of ^{64}Cu to copper-binding enzymes in the liver [51] as well as to serum proteins, which causes higher blood pool activity and background signal [43]. Thus, other ^{64}Cu chelating agents, such as 1,4,8,11-tetraazacyclotetradecane-1,4,8,11-tetraacetic acid (TETA) and 4,11-bis(carboxymethyl)-1,4,8,11-tetraazabicyclo[6.6.2] hexadecane (CB-TE2A), offer better stability and have shown effectiveness in vivo. Recently, Fani and colleagues demonstrated the effectiveness of the triaza chelating agent 1,4,7-triazacyclononane,1-glutaric acid-4,7-acetic acid (NODAGA), which rapidly chelates ^{64}Cu at room temperature and exhibits excellent in vivo stability [52]. In their work, peptides were ^{64}Cu-labeled and produced excellent tumor visualization at 1 hr with low background signal. This result was a significant advancement in the development of ^{64}Cu radiotracers since peptides labeled with DOTA generally require delayed (i.e., next day) imaging to allow for clearance of background signal in order to obtain sufficient contrast. Most importantly, their findings showed the attractiveness of NODAGA as an efficient, biocompatible, and stable chelation approach that could potentially be applied to all classes of imaging agents, including nanoparticles, to enable imaging shortly after injection for longitudinal assessment of biodistribution and PK. Future studies examining the imaging properties of NODAGA with different biomolecular constructs will be critical in assessing its utility as a chelator for radiotracer development.

2.4.4 FLUORINE-18

[18]F is a cyclotron-produced, positron-emitting isotope that is frequently used to label biomolecules for PET [53]. One of the key attributes that makes [18]F a significant radionuclide for molecular imaging is its low positron energy, which provides the highest possible resolution in a PET camera. In addition, ^{18}F has a 109.8 min half-life that allows sufficient time for complex radiosyntheses to be performed that, in most cases, are carried out using automated modules. While mostly known for its role

in metabolic imaging with ^{18}F-FDG, ^{18}F-labeled compounds have also been developed for imaging applications that include cellular proliferation [54], tumor hypoxia [55], ER detection [56], reporter gene expression [57,58], and neurological disorders [59,60]. [18]F has most often been associated with small molecules where site-specific conjugation can be achieved through electrophilic addition onto an aromatic ring or nucleophilic substitution involving a leaving group. However, advances in bioconjugation techniques have provided reagents that can be radiolabeled with ^{18}F and efficiently incorporated into nanoparticles through surface modifications, thus encouraging expanded use of this radiohalogen for nanomedicine.

2.4.5 OTHER RADIONUCLIDES

Radioactive forms of iodine play a significant role in nuclear medicine due to diverse emission properties and applications for diagnosis and therapy. Of particular significance is the fact that radioactive iodine is considered a nonresidualizing label since the metabolites (e.g., iodide or iodotyrosines) of radioiodinated proteins are quickly released from the cells and excreted via the kidneys, whereas metabolites of radiometal-chelated proteins (residualizing) are trapped within cells, leading to increased retention of activity over time [61]. This property results in lower background signal and improved contrast, and is a major advantage of radiotracers that employ radioiodine labels. For nanoparticles, ^{125}I has been employed on the basis of its long half-life (60.1 d), which is suitable for long-term tracking of agent biodistribution, and the ability to measure its low energy (35 keV) emissions in excised tissues. The application of radioactive iodine for PET is also possible through the use of [124]I, an emerging radionuclide attracting increasing interest for long-term clinical and small-animal PET studies [62]. [124]I has a long half-life (4.2 d) that is appropriate for studying the PK properties of long-circulating compounds through delayed imaging. Labeling with radioactive iodine can occur by direct attachment of the radionuclide to the side chain of tyrosine residues on peptides/proteins or to nanoparticles that have been conjugated to protein-based targeting moieties, adsorption or loading into the internal compartment of a nanoparticle, or insertion into polymers or other units used for surface functionalization.

Gallium-67 (^{67}Ga) is another SPECT radionuclide used in the study of radiolabeled nanoparticles and possesses a 3.3 d half-life that is compatible with the needs for delayed imaging. However, despite its clinical role for imaging infections, lymphomas, and granulomatous diseases [63,64], its use with molecular imaging agents is infrequent owing to its decay through a broad range of γ-ray emissions that result in poor image quality [65].

Zirconium-89 ([89]Zr) is a long-lived PET radionuclide ($t_{1/2} = 3.3$ d) that has gained attention in recent years for antibody imaging. The fact that [89]Zr forms a stable chelate with desferrioxamine B (Df) and provides PET images that can be used for quantitative analysis of uptake is a key advantage of its use with antibodies as well as nanoparticles. Moreover, [89]Zr-based PET imaging has been investigated preclinically for a wide variety of cancer-related targets [66] as well as in recent clinical studies that were successful for detecting HER2-positive lesions in patients with metastatic breast cancer [67] and lymph node metastases in patients with head and neck cancer

[68,69]. Although the presence of a high-energy gamma emission is a disadvantage of ^{89}Zr, which may limit the radioactive dose that can be administered into patients, a growing infrastructure consisting of a more readily available radioisotope supply and well-developed radiochemistry using commercially available chelating agents suggest a bright future for radiotracer development with this positron emitter.

2.5 IMAGING WITH RADIOLABELED NANOPARTICLES

2.5.1 LIPOSOMES

Liposomes are a type of drug delivery vehicle consisting of spherical particles formed by a lipid bilayer that encloses an aqueous compartment. They are attractive for drug delivery due to their biodegradability, ease of preparation, lack of toxicity, and ability to tune biodistribution based on size, charge, and lipid composition. In nuclear imaging, liposomes have a long history of use that dates back to the 1970s [70–72] and includes early clinical studies in cancer patients [73,74]. During their manufacturing process, different techniques are used to radiolabel liposomes and are divided into four general groups: passive encapsulation, membrane labeling, surface chelation, and remote loading (Figure 2.6). The most common approaches, surface chelation and remote loading, benefit from the use of chelation strategies that have been well studied with conventional radiotracers and offer better yields, simpler purification techniques, and higher stability.

2.5.1.1 Single-Photon Emitting Agents

The majority of reports on radiolabeled liposomes involve SPECT radionuclides, for which multiple labeling methods have been summarized in reviews by Philips [75] and Petersen et al. [76]. Among these, the most effective techniques use remote loading of radioactivity into preformed liposomes. This is generally performed through the use of a lipophilic chelating agent that binds a radiometal and acts as a carrier to deliver the radioactive complex into the liposome, where further processing occurs and trapping is achieved. In this section, we focus on the most prominently used SPECT emitters, ^{99m}Tc and ^{111}In, to highlight the development of radiolabeled liposomes and their nuclear imaging applications.

In the case of ^{99m}Tc, many chelating agents and approaches are available for radiolabeling, including the lipophilic chelating agent hexamethylpropyleneamine oxime (HMPAO). The proposed mechanism in this approach involves HMPAO chelation of ^{99m}Tc and subsequent delivery of the radioactive complex inside liposomes, where preloaded glutathione (GSH) acts as a reducing agent to form a hydrophilic complex that becomes trapped within the liposomes. The ^{99m}Tc-HMPAO trapping technique has been applied to multiple liposome-encapsulated compounds to examine imaging and biodistribution in animal models and has proved to be a major advance in the field of liposome imaging. Studies conducted on hemoglobin-encapsulated liposomes effectively identified the interaction of the imaging agent with the organs of the RES (namely, liver and spleen) and demonstrated that the encapsulation of the label within the liposome retards the metabolism of ^{99m}Tc-HMPAO compared to previously observed instability of the chelation complex in saline [77]. Further

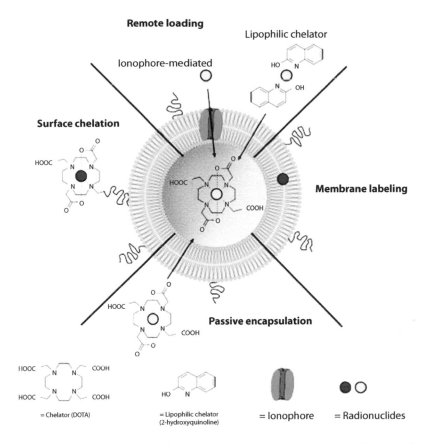

FIGURE 2.6 **(See color insert.)** Schematic diagram of the remote loading, membrane labeling, passive encapsulation, and surface chelation methods for preparing radioactive liposomes. Radionuclides can be associated with the lipid membrane by hydrophobic interaction, through membrane conjugation, or surface chelation using chelator–lipid conjugates in preformed liposomes (*blue radionuclides*). Radionuclides can alternatively be encapsulated inside liposomes during lipid hydration or can be transported through the lipid membrane of preformed liposomes by ionophores or lipophilic chelators (*yellow radionuclides*). In the latter case, the radionuclides are trapped inside the aqueous lumen by a hydrophilic chelator with high affinity for the radionuclide. (Reproduced with permission from Petersen, A. L. et al. 2012. *Advanced Drug Delivery Reviews* 13:1417–1435.)

characterization of this approach involved comparison of neutral versus negative surface charge on the liposome and found comparable tumor targeting, but showed that neutral preparations had lower RES accumulation [78].

To further reduce the uptake of radiolabeled liposomes by phagocytic cells of the RES and improve imaging performance, new ^{99m}Tc liposome formulations were developed using PEG to prolong circulation time and enhance uptake by target tissues. A study by Tilcock et al. examined the in vivo properties of ^{99m}Tc PEGylated liposomes and showed a dramatic increase in blood circulation time compared to non-PEGylated liposomes [79]. Oyen and colleagues also investigated the performance of these [99m]Tc "stealth" liposomes for detecting infection and inflammation in rats and found preferential accumulation in abscesses that resulted in very high target-to-nontarget ratios, which were attributed to the extended circulation of the liposomes [80]. A combinatorial approach using ^{99m}Tc-HMPAO-loaded Doxil, a PEGylated liposomal formulation of doxorubicin that is FDA approved for cancer treatment, was also tested and yielded a theranostic compound that could accurately estimate doxorubicin concentration in a murine tumor model [81].

Despite the effectiveness of HMPAO for ^{99m}Tc loading, other chelation approaches have been examined to improve the efficiency of liposome labeling and purification. Hnatowich et al. employed a chelation strategy where the standard ^{99m}Tc chelator, DTPA, was incorporated into liposomes during their production and resulted in excellent labeling yields [82]. Using a different chelation scheme, Laverman et al. incorporated a derivative of the ^{99m}Tc chelator hydrazinonicotinamide (HYNIC) into the bilayer of PEG liposomes and were also able to demonstrate rapid, one-step labeling with high yields that do not require purification, making their method attractive for routine agent formulation [83]. The HYNIC approach provided a major advancement in the development of radiolabeled liposomes and has been used for detection of intra-abdominal abscesses [84] and focal infections [85], visualization of early adhesion formation after experimental peritonitis [86], and assessment of the impact of lipid dose on PK and biodistribution in animal models as well as patients [87].

[111]In is another single-photon emitter that has been widely used for liposomal labeling. The use of ^{111}In as a label is attractive because it gives the ability to perform delayed imaging beyond the 24 hr time window for which ^{99m}Tc use is feasible, and can be detected for several days postinjection. By using this longer lived radionuclide, imaging studies can be carried out over several days to assess PK and biodistribution properties, which may not be clear at earlier time points. As with ^{99m}Tc, multiple approaches have been developed to label liposomes with [111]In. The most effective technique is through remote loading of radioactivity into preformed liposomes and can be achieved by two different methods. The first method involves the use of the lipophilic compound 8-hydroxiquinoline (known as oxine), which has been widely studied as it permits chelation of ^{111}In and diffusion into cells, where it then undergoes dissociation and leaves ^{111}In available for interaction with other intracellular components or internal chelating agents. In a study by Hwang et al., oxine was used to chelate ^{111}In for transport through the lipid bilayer to the encapsulated chelating agent, nitrilotriacetic acid (NTA) [88]. Labeling efficiencies of 90% were achieved and marked a significant improvement over methods that incorporated ^{111}In-oxine into the lipid bilayer during liposome production. Similar approaches have been described using acetylacetone as a water-soluble lipophilic chelate for loading ^{111}In into liposomes [89], or DTPA as a strong chelating agent encapsulated within the aqueous compartment to trap incoming [111]In [90].

The second approach for remote loading is to insert an ionophore in the lipid membrane of a liposome to enable transit of radiometals across the lipid bilayer, where they can be sequestered by an encapsulated chelating agent. Ionophore A23187 has been used for this purpose with ^{111}In-labeled phospholipid vesicles [91] and resulted in the development of Vescan, a liposomal agent used for cancer imaging in patients

[92–95]. These trials represented some of the earliest clinical experiences with liposomes and showed that nanoparticles were well tolerated with radiation doses comparable to existing radionuclide techniques. Although Vescan underwent testing in phase III clinical trials, it was not commercialized. However, the experience gained throughout the development process helped establish a paradigm for small, stable liposomes, key manufacturing and quality control technology, regulatory pathways for clinical use of liposomes, and essential safety information on injectable small unilamellar vesicles [96].

Similarly to ^{99m}Tc-liposomes, several groups have developed chelator-derivatized phospholipid building blocks to equip the surface of liposomes with chelating moieties to facilitate ^{111}In labeling. Helbok et al. prepared a DTPA-derivatized lipid-based liposome that possessed >90% labeling yield for several different radiometals and showed in vivo imaging properties that suggest utility as a multifunctional nanoparticle for targeting applications [97]. Most importantly, their surface labeling approach resulted in an ^{111}In-agent, which proved highly stable when challenged by excess DTPA at 24 hr incubation and showed the anticipated RES uptake with high blood levels. A similar study was recently published using the macrocyclic chelator DOTA, which is superior to DTPA for many labeling applications [98]. DOTA was attached to the lipid head group and labeled with Gd, ^{64}Cu, or ^{111}In to produce a multifunctional liposomal formulation with significant potential for various diagnostic imaging applications (Figure 2.7).

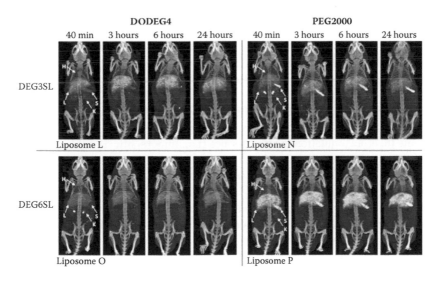

FIGURE 2.7   (See color insert.) SPECT/CT images showing that both of the control PEG2000 liposomes (N and P) appear to show a characteristically high uptake in the spleen compared to both DODEG4 liposomes (left) that possess a lipid multifunctional, multimodal shielded liposomal formulation. Images were acquired over 24 hr (H—heart, L—liver, S—spleen, and K—kidney). (Reproduced with permission from Mitchell, N. et al. 2013. *Biomaterials* 4:1179–1192.)

2.5.1.2   Positron Emitting Agents

Although SPECT has predominantly been used for imaging radiolabeled liposomes, the clinical importance of PET, growing use of hybrid preclinical scanners, and increased availability of radionuclides such as ^{64}Cu have led to a rise in interest for PET-based liposomal agents. Moreover, advances in ^{64}Cu chelation chemistry have introduced several highly stable chelation approaches that have improved in vivo imaging properties of conventional radiotracers, and can be applied for liposome labeling with the likelihood of high yields and stability. A study by Seo et al. showed that a ^{64}Cu specific chelator, 6-[p-(bromoacetamido)benzyl]-1,4,8,11-tetraazacyclotetradecane-N,N',N'',N'''-tetraacetic acid (BAT), could be conjugated with an artificial lipid to form a BAT–PEG lipid and provide radiolabeling yields as high as 95% with excellent stability [99]. In another investigation, this approach was applied in a mouse model of ductal carcinoma in situ to estimate the tumor vascular volume and permeability by PET [100]. The authors found that the radiolabeled particles were effective for detecting the transition from premalignant to malignant lesions and effectively quantified the associated changes in vascular permeability.

^{18}F-based approaches have also been investigated for liposomal labeling. Oku and colleagues used ^{18}F-FDG encapsulation as a means to track liposomes noninvasively [101,102]. Due to inefficient labeling with ^{18}F-FDG, Marik et al. synthesized a radiolabeled diglyceride, 3-[^{18}F]fluoro-1,2-dipalmitoylglycerol (^{18}F-FDP) for radiolabeling long-circulating liposomes that, contrary to encapsulation of ^{18}F-FDG, were incorporated into the phospholipid bilayer [103]. The in vivo imaging data showed that the long-circulating liposomes remained in the bloodstream for at least 90 min and that the free ^{18}F-FDP was not metabolized in the myocardium. Urakami et al. also focused on new methodology referred to as the "solid-phase transition method" for rapid labeling of preformed liposomes and were able to achieve high incorporation yields and identify a strategy for monitoring the behavior of liposomes with varying size [104].

2.5.1.3   Actively Targeted Liposomal Imaging Agents

Thus far, the radiolabeled liposomes discussed have been passively targeted and attribute their accumulation in target tissues (i.e., tumors, infections) to leaky vasculature and dysfunctional lymphatics in what is known as the enhanced permeability and retention (EPR) effect. While an increased concentration of a radiolabeled liposome can occur within target tissue via the EPR effect, molecular specificity is still lacking. Methods for surface modification of liposomes using PEGylation and chelating moieties provided the foundation to rationally introduce ligands with affinity for certain biomarkers that have diagnostic relevance in different disease models. A number of liposomal targeting strategies have been described and include vascular targets, cell surface receptors, extracellular matrix proteins, and others [76]. In the targeting approach by ElBayoumi and Torchilin, the anticancer monoclonal antibody 2C5 (mAb 2C5) was added to doxorubicin-loaded long-circulating liposomes in order to increase tumor-specific uptake and subsequently deliver a larger drug payload [105]. Following ^{111}In labeling, imaging and biodistribution studies revealed higher tumor accumulation of mAb 2C5-functionalized liposomes compared to non-targeted control agents and enabled accurate prediction of the therapeutic efficacy

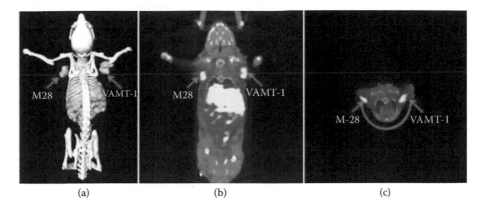

(a)   (b)   (c)

FIGURE 2.8 **(See color insert.)** SPECT/CT fused image of ^{111}In-IL-M1$_{(50)}$ taken 24 h after injection; (a) three-dimensional reconstruction, (b) coronal view, (c) transverse view. The uptake of ^{111}In-IL-M1$_{(50)}$ in both epithelioid (M28) and sarcomatoid (VAMT-1) mesothelioma tumors at 24 hr was shown. (Reproduced with permission from Iyer, A. K. et al. 2011. *Biomaterials* 10:2605–2613.)

of the functionalized liposomes. From this work, the authors were able to show that molecular targeting can enhance tumor uptake beyond the EPR effect.

Iyer et al. selected a human single-chain antibody fragment—an emerging approach with significant potential in molecular imaging—to target ^{111}In-liposomes to epithelioid and sarcomatoid mesothelioma [106]. The authors showed tumor detection in vivo (Figure 2.8) suggesting that liposomes could be effectively functionalized with small protein fragments and exhibit improved affinity for cancer detection compared to the EPR effect.

Peptide-targeted liposomes have also been described using SST analogs such as octreotide and octreotate. In a study by Petersen and colleagues, the attachment of octreotate to liposomes loaded with ^{64}Cu resulted in faster initial blood clearance in comparison to control liposomes and had significantly higher tumor/muscle ratios [107]. Similar findings were reported by Helbok et al. with the SST analogue tyrosine-3-octreotide in mice with tumor xenografts [108]. These studies are critical for gaining insight into the feasibility of peptide-targeted liposomes and may guide future efforts to optimize agent design. In addition, the selection of clinically used imaging models, such as SST receptor targeting, provides a validated benchmark against which liposome-based constructs can be evaluated and enhances the agent characterization process.

2.5.2 Carbon Nanotubes

Various carbon-based nanoparticles have been studied as imaging agents, including carbon nanotubes (CNTs), C$_{60}$ fullerenes, perfluorocarbon nanoemulsions, and graphene oxide particles [109]. Among these, carbon nanotubes have emerged as an attractive delivery platform for biomedical applications. CNTs are members of the fullerene structural family and consist of graphite sheets rolled at specific and discrete angles. CNTs can be described as being either single-walled CNTs (SWCNTs) or multiwalled CNTs (MWCNTs) and possess unique properties that have led to their use in various fields, including biomedical research [110]. Early investigations with radiolabeled CNTs were focused on single-photon emitters using various approaches for radiolabeling. A report from Wang and co-workers showed for the first time the quantitative analysis of SWCNT accumulation in animal tissues using [125]I [111]. The hydroxylated SWCNTs were taken up in the stomach, kidneys, and bone, but had slow whole-body clearance with persistent signal from bone and kidneys for up to 6 d. Mackeyev et al. performed in vitro analysis of the adsorption of ^{125}I in SWCNTs in comparison with other carbon-based materials such as fluorinated SWCNTs, cut SWCNTs, charcoal, graphite, fluorinated graphite, and C_{60} [112]. Cut SWCNTs, uncut SWCNTs, and charcoal all showed highly efficient adsorption of ^{125}I, while cut SWCNTs demonstrated the best retention of the radionuclide. A different labeling approach was shown by Singh and colleagues, who employed ^{111}In as a single-photon emitter for labeling [113]. They functionalized the surface of SWCNTs with DTPA to permit efficient labeling of ^{111}In and studied in vivo properties in mice. Gamma scintigraphy indicated that the functionalized SWCNTs were not retained in any of the RES organs and were rapidly cleared from blood through the renal excretion route.

To examine SWCNTs by PET, McDevitt and co-workers employed DOTA surface functionalization to chelate [86]Y and compared tissue biodistribution and PK to an analogous ^{111}In construct modified with DTPA [114]. PET imaging of [86]Y-DOTA-SWCNTs revealed blood clearance within 3 hr with primary distribution to the kidneys, liver, spleen, and bone. Although the activity that accumulated in the kidney cleared with time, the whole-body clearance was slow with this agent.

In order to reduce the cytotoxicity and water insolubility of pristine SWCNTs, Hong et al. described a mild strategy for surface modification based on azomethine ylide 1,3-dipolar cycloaddition [115]. The authors selected carbohydrate-containing structures, as they are widely seen in nature and play essential roles in many biological processes, and performed surface glycosylation with N-acetylglucosamine (GlcNAc). With SPECT/CT imaging, prominent tracer accumulation was seen in the lungs and was not visible in the thyroid or stomach, suggesting minimal leakage of ^{125}I from the nanoparticles and affording a low background signal for ultrasensitive imaging (Figure 2.9).

To further improve biological applications, various targeting strategies have been applied with SWCNTs. To capitalize on the effectiveness of integrin targeting with conventional radiotracers, Liu et al. selected the cyclic integrin-targeting peptide RGD (arginylglycylaspartic acid) for vascular targeting and observed high tumor uptake of functionalized SWCNTs in mice (Figure 2.10), which they attributed to the multivalent effect of the nanoparticles [116]. An antibody-based SWCNT construct for vascular targeting was developed by Ruggerio and collaborators with the E4G10 antibody targeted against the VE-cad epitope expressed in the tumor angiogenic vessels [117]. The authors showed that antibody–SWCNT conjugates had favorable blood clearance kinetics and were found to be well tolerated and safe in animal models. Utility for PET imaging and tumor treatment was also demonstrated using diagnostic and therapeutic radionuclide labeling strategies. Tumor cells have also been targeted by an SWCNT construct attached to Rituximab, a chimeric monoclonal

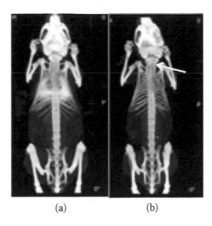

(a) (b)

FIGURE 2.9 **(See color insert.)** Whole-body SPECT/CT imaging in mice at 7 d postinjection showing (a) persistent lung accumulation of GlcNAc^D–Na^{125}I@SWNTs and (b) thyroid accumulation of free Na^{125}I (unencapsulated), indicating effective and complete encapsulation of radionuclide within the nanocapsule and long-term stability of the construct. (Adapted with permission from Hong, S. Y. et al. 2010. *Nature Materials* 6:485–490.)

antibody against the protein CD20, to study lymphoma in mice [118]. Using DOTA-mediated chelation of ^{111}In, the targeted SWCNTs exhibited comparable uptake to ^{111}In-Rituximab in vitro, although in vivo binding was markedly lower. The authors attributed this result to a possible mixture of appended antibodies with different geometric distributions on the surface of SWCNTs, thus impacting effectiveness for epitope targeting. In addition, their strategy to modify randomly distributed lysine residues on the antibody for maleimide linkage could have interfered with key residues within the pharmacophore and further affected binding. Nonetheless, they suggested further studies to optimize the imaging properties of their prototype constructs through the use of the high aspect ratio of SWCNTs.

MWCNTs have also been evaluated using nuclear imaging techniques. Guo et al. used glycosylation to enhance the solubility of MWCNTs and conducted the first in vivo assessment of distribution in mice [119]. After ^{99m}Tc labeling, SPECT imaging revealed that the functionalized MWCNTs moved easily among the compartments and tissues of the body, behaving like active molecules. The authors did not observe any severe acute toxicity and implicated improved biocompatibility after functionalization. More recently, the use of radiopharmaceutical chemistry and noninvasive imaging have permitted examination of the biodistribution and PK of a novel therapeutic MWCNT conjugate through ^{99m}Tc labeling [120], as well as assessment of the effects of surface chemical modification on organ distribution and clearance of MWCNTs with ^{111}In [121].

2.5.3 Dendrimers

Dendrimers are hyperbranched, artificial macromolecules that possess three basic architectural components: a core, an interior of shells consisting of repeating branch-cell units, and terminal functional groups. Their step-wise synthesis is well controlled

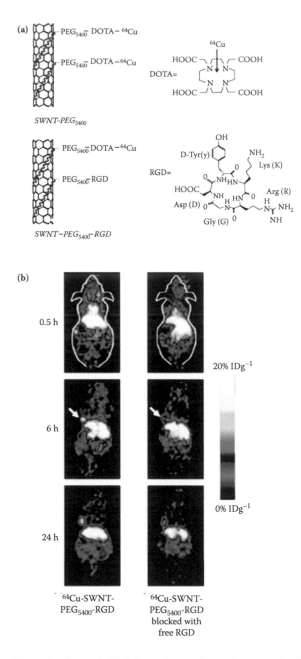

FIGURE 2.10   (See color insert.) (a) Schematic drawings of noncovalently functionalized SWNT-PEG$_{5400}$-RGD conjugates with ^{64}Cu-DOTA. The hydrophobic carbon chains (blue segments) of the phospholipids strongly bind to the sidewalls of the SWNTs, and the PEG chains render water solubility to the SWNTs. (b) MicroPET images of U87MG tumor-bearing mice showing high tumor uptake of ^{64}Cu-SWNT-PEG$_{5400}$-RGD, which was significantly reduced by co-injection of free RGD peptide. (Adapted with permission from Liu, Z. et al. 2007. *Nature Nanotechnology* 1:47–52.)

and provides molecules with defined molecular weights and tunable properties such as size, polarity, and solubility [122]. This makes dendrimers highly attractive for drug development as chemists can utilize these quantized building blocks for the synthesis of well-defined, more complex nanostructures. Moreover, they possess highly reactive pendant functional groups on their surface that can be used for covalent conjugation of targeting and diagnostic moieties, making them suitable for drug delivery and imaging. The first report on radiolabeled dendrimers was provided by Roberts et al. and involved the use of starburst dendrimers as intermediate linker molecules for preparation of antibody–drug conjugates [123]. Using a porphyrin-based chelation approach, the authors showed that radiolabeled antibodies could be prepared with higher specific activity using dendrimer-containing conjugates. From these pilot studies, the authors demonstrated the utility of dendrimers for antibody modification and emphasized key advantages of their approach, such as control over molecular weight, number of terminal functional groups, and homogeneity of composition, that are not possible with commonly used polymer-based linkers.

Among the different dendrimer scaffolds used for nuclear imaging, poly(amidoamine) (PAMAM) dendrimers have undergone the most extensive characterization. PAMAM dendrimers possess a large number of surface amino groups that correlate to the generation of the dendrimer construct (i.e., generation-1 (G1) has 8 available sites, generation-2 (G2) has 16 available sites, etc.) and allow multiple chelating moieties to be attached, thus improving labeling kinetics and specific activities (Figure 2.11) [124]. Kobayashi and colleagues prepared a high specific activity ^{111}In-antibody conjugate targeting murine alkaline phosphatase and consisting of 43 molecules of a DTPA derivative attached to a PAMAM (G4) containing 64 amines [125]. Their approach effectively maintained the immunoreactivity of the antibody and had more rapid systemic clearance than a conventionally prepared conjugate. In a multimodality application, Criscione and collaborators used PAMAM (G4) dendrimers for the preparation of a dendrimer-based agent consisting of triiodinated CT contrast and ^{99m}Tc–DPTA coordination complexes for SPECT/CT imaging [126]. The authors successfully developed a long circulating blood-pool imaging agent that showed colocalization of CT and nuclear signals in mice (Figure 2.12). They also pointed out how simple synthetic modifications could be introduced into their agent to permit labeling with longer lived SPECT or PET radionuclides to maximize imaging capabilities.

Alternative dendrimer types have also been developed for use with radiolabeled imaging agents. For example, to construct a highly safe carrier molecule with lower reported toxicity than PAMAM, Okuda and collaborators designed a lysine (G6) dendrimer (KG6) comprising L-lysine branch units and evaluated in vivo properties by labeling with [111]In [127]. Biodistribution studies revealed rapid clearance from the blood stream and nonspecific accumulation in the liver and kidney. By introducing different levels of PEGylation—modification rates of approximately 7.8% (PEG(10)-KG6) and 59.4% (PEG(76)-KG6)—increased blood retention was observed as well as gradual tumor accumulation of PEG(76)-KG6 as a result of the EPR effect. Agashe and colleagues studied the effects of carbohydrate (mannose and lactose) coating on a poly(propylene imine) (G5) dendrimer in mice and found high labeling efficiency and stability of their dendrimer formulations [128]. Mannose

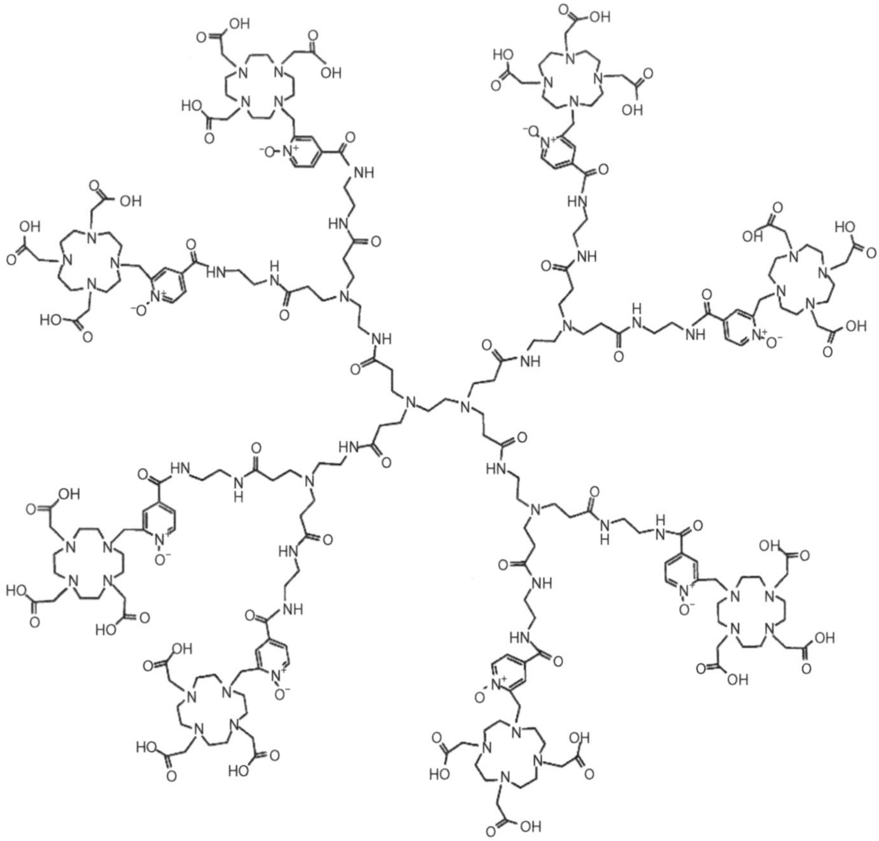

FIGURE 2.11 Chemical structure of a G1 dendrimer scaffold modified with eight chelators. (Reproduced with permission from Biricova, V. et al. 2011. *Journal of Pharmaceutical and Biomedical Analysis* 3:505–512.)

coated dendrimers exhibited the fastest clearance rate from blood, though significant renal uptake occurred. The authors observed high liver uptake for all dendrimeric polymers and attributed the significant retention of carbohydrate-coated agents at 6 hr after injection to lectin–carbohydrate interaction within liver cells.

Targeting with dendrimeric complexes has also been achieved using various approaches. Dijkgraaf et al. used Cu-catalyzed click chemistry (a method tailored to generate substances quickly and reliably by joining small units together) to form a series of $\alpha_v\beta_3$ integrin-directed monomeric, dimeric, and tetrameric RGD dendrimers for tumor targeting (Figure 2.13) [129]. Enhanced receptor affinity was observed in vitro with multivalent RGD dendrimers compared to the monomeric derivative. In vivo studies found that ^{111}In-labeled DOTA-RGD dendrimers showed specifically enhanced uptake in $\alpha_v\beta_3$ integrin-expressing tumors, with tetrameric complexes having better tumor targeting properties than dimeric and monomeric formulations. Folate-targeted dendrimers have also been described by Zhang and

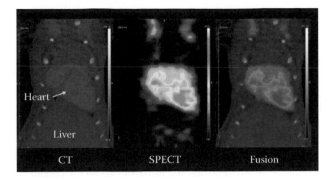

FIGURE 2.12 (See color insert.) Multimodal imaging of the G4-[[[[Ac]-TIBA]-DTPA]-mPEG₁₂] dendrimer construct in normal mice by microCT and microSPECT. The fusion image shows significant colocalization of the nuclear and x-ray contrast components. (Reproduced with permission from Criscione, J. M. et al. 2011. *Bioconjugate Chemistry* 9:1784–1792.)

FIGURE 2.13 Schematic representation of tetrameric RGD dendrimers developed using "click chemistry."

colleagues using the PAMAM (G5) scaffold [130]. By incorporating DTPA moieties into the dendrimer construct, ^{99m}Tc labeling was achieved with yields > 95% and showed excellent stability with more than 80% of the agent remaining intact in both in vitro and in vivo assays. Compounds containing PEG linkers demonstrated higher in vitro uptake in KB cells than those without PEG, suggesting that indirect folate conjugation through the PEG spacer could enhance receptor binding. The findings were confirmed in vivo in mice xenografts and encouraged further examination of molecule design to optimize the dendrimer-based ^{99m}Tc–folate conjugate for cancer diagnosis.

Building on prior work with highly stable and specific peptide nucleic acid (PNA) hybridization, Amirkhanov and others developed ^{111}In-labeled nanoparticles to target *KRAS2* mRNA for imaging of human pancreatic cancer xenografts [131]. Polydiamidopropanoyl (PDAP) dendrimers were used and contained a short, cyclized insulin-like growth factor 1 (IGF-1) peptide analog, D(Cys-Ser-Lys-Cys), to permit receptor-mediated endocytosis of PNA probes into malignant cells that overexpress IGF-1 receptors. Their imaging studies revealed that the ^{111}In-labeled KRAS2 PNA-IGF-1 nanoparticles could permeate xenografts and accumulate in a sequence-specific manner. Importantly, imaging of AsPC1 xenografts with fully matched probes resulted in an increase in tumor/muscle ratio that correlated to increased size of the dendrimer and the number of DOTA chelating groups (up to $n = 16$). Such imaging characteristics were not shown with single mismatch probes, regardless of the size and abundance of ^{111}In-DOTA complexes, and demonstrated the ability of the PNA-targeted nanoparticles to achieve mRNA targeting with single mismatch specificity.

2.5.4 IRON OXIDE NANOPARTICLES

Iron oxide (IO) particles are a class of nanoparticles which consist of a crystalline magnetite structure and have a long history of biomedical use [132]. Among these, superparamagnetic iron oxide (SPIO) particles are among the most successful examples of nanoparticles available and have shown widespread clinical utility as MRI contrast agents [133–135]. Attractive characteristics such as low toxicity, decomposition within cells, and biocompatibility make IO-based contrast agents particularly well suited for biomedical applications. Furthermore, they can be readily functionalized through different surface chemistry approaches to achieve specific targeting for molecular imaging applications with multiple modalities. For example, Glaus and colleagues developed bifunctional SPIO nanoparticles that were coated with PEGylated phospholipids and contained DOTA moieties for PET [136]. ^{64}Cu-labeling was performed with high yield and purity, and stability in mouse serum was confirmed after 24 hr incubation. Measurement of PK by PET and organ biodistribution in mice revealed early blood pool activity, which diminished significantly at 4 hr, while high uptake occurred in the organs of the RES. PET imaging correlated well with biodistribution findings and indicated ^{64}Cu-SPIO signals primarily in the heart (suggesting extended blood retention), liver, and spleen at 24 hr postinjection. Gastrointestinal (GI) uptake was also noted and attributed to biliary excretion of ^{64}Cu-SPIOs (and their metabolic products) that were sequestered in the liver at earlier time points.

In a study by Lee et al., a targeted IO-based imaging agent was synthesized for multimodal PET/MRI imaging of integrin $\alpha_v\beta_3$ expression [137]. To overcome the multistep process associated with surface functionalization of IO particles coated with dextran, PEG, and other substances, the authors selected polyaspartic acid (PASP) to provide a simpler method for attaching RGD for tumor targeting and DOTA for ^{64}Cu radiolabeling. Multimodal small-animal imaging was performed with PET and MRI and revealed integrin-specific uptake of RGD-PASP-IO nanoparticles and prominent uptake in the RES.

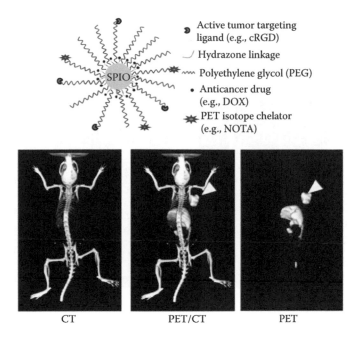

FIGURE 2.14 Top: illustration of multifunctional RGD-conjugated SPIO nanocarriers for combined tumor-targeting drug delivery and PET/MR imaging. Bottom: representative PET/CT images of a U87MG tumor-bearing mouse at 6 hr postinjection of ^{64}Cu-labeled RGD-conjugated SPIO nanocarriers (tumor indicated by arrow). (Adapted with permission from Yang, X. et al. 2011. *Biomaterials* 17:4151–4160.)

Yang and others investigated a theranostic approach by using integrin targeting as an approach to deliver chemotherapy with PEGylated IO nanoparticles [138]. Their construct consisted of pH-sensitive hydrazone bonds to achieve pH-responsive release of doxorubicin, RGD peptides for vascular targeting, and the macrocyclic chelating agent 1,4,7-triazacyclononane N, N', N''-triacetic acid (NOTA) for ^{64}Cu labeling. Receptor binding was confirmed with U87MG cells and increased cytotoxicity was observed with RGD-conjugated constructs. The authors also showed that their surface functionalization tactics did not have a significant impact on the effectiveness of the SPIO nanocarriers to serve as MRI contrast agents. Furthermore, PET and tissue biodistribution in U87MG tumor-bearing mice demonstrated the effectiveness of targeting as RGD-targeted nanoparticles exhibited preferential tumor accumulation compared to nontargeted controls. A representative PET/CT image is shown in Figure 2.14.

More recently, Yang et al. examined the in vivo properties of IO nanoparticles that were combined with gold nanoparticles into a multifunctional hetero-nanostructure useful for PET, MRI, and optical imaging [139]. By conjugating an antiepidermal growth factor receptor (EGFR) affibody and labeling with ^{64}Cu, PET imaging in mice successfully demonstrated tumor-specific uptake that correlated with MRI findings and confocal optical imaging, suggesting the suitability of the nanoparticle composition for multimodality imaging.

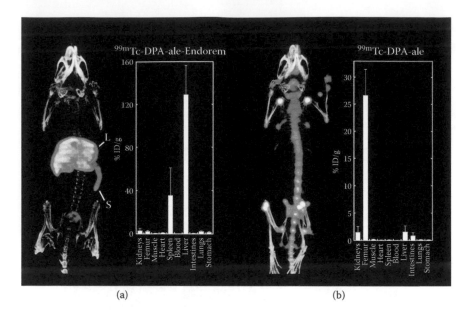

FIGURE 2.15 **(See color insert.)** Whole-body SPECT/CT images and biodistribution studies of (a) a ^{99m}Tc-labeled SPIO-bisphosphonate construct and (b) a conventional ^{99m}Tc-bisphosphonate agent. (Reproduced with permission from Torres Martin de Rosales, R. et al. 2011. *Bioconjugate Chemistry* 3:455–465.)

Studies have also been conducted with SPECT by attaching ^{99m}Tc-labeled bisphosphonate directly to the surface of the IO core [140] or with a PEG polymer conjugate containing a bisphosphonate group bound to the surface of ultrasmall SPIO nanomaterials (USPIOs) [141]. In both studies, RES uptake resulted from conjugation to the nanoparticles (Figure 2.15) and favorable imaging properties were shown for SPECT and MRI. In another study, Madru and colleagues directly labeled the surface of PEGylated SPIO nanoparticles with ^{99m}Tc with high efficiency and were able to identify the sentinel lymph node in rats by SPECT/MRI [142]. These studies described strategies for ^{99m}Tc labeling, which could be modified for use with other SPECT radionuclides, and demonstrated the utility of SPECT-based approaches for multimodality imaging with IO nanomaterials.

2.5.5 QUANTUM DOTS

Colloidal semiconductor quantum dots (QDs) are single crystals a few nanometers in diameter whose size and shape can be precisely controlled by the duration, temperature, and ligand molecules used during synthesis [143]. QDs have desirable properties of high quantum yield, resistance to photobleaching, narrow emission peak, and tunable emission wavelength—making them attractive for molecular imaging applications [144–146]. While initial in vivo studies were based on nontargeted QDs, more recent studies have employed targeting moieties to achieve specificity and surface conjugation to improve solubility and biocompatibility for in vivo use. In addition, several groups have added a radioactive label to produce a multimodal construct that can be used for nuclear/optical imaging while also providing direct quantification of in vivo distribution. A study by Cai and colleagues described the use of a dual-function PET/near infrared fluorescence (NIRF) probe containing ^{64}Cu to allow for in vivo assessment of the PK and targeting properties of QDs [147]. The QDs underwent surface modification to introduce amine groups that were used to conjugate DOTA for radiometal complexation and RGD for integrin $\alpha_v\beta_3$-targeted PET/NIRF imaging. Using U87MG human glioblastoma cells, in vitro assays revealed integrin $\alpha_v\beta_3$-specific binding of DOTA-QD-RGD, indicating that these nanomaterials could be successfully targeted. In tumor-bearing mice, ^{64}Cu-DOTA-QD-RGD demonstrated significantly higher tumor uptake than the nontargeted ^{64}Cu-DOTA-QD control agent, with tumor/muscle ratios that were four-fold higher. In vivo imaging by PET and NIRF showed excellent correlation and histological analysis provided evidence of agent localization primarily in the tumor vasculature.

In a subsequent study, the authors applied the same animal model to evaluate a ^{64}Cu-labeled QD protein conjugate for multimodal PET/NIRF imaging of vascular endothelial growth factor receptor (VEGFR) expression [148]. They showed that the QD-VEGF nanoconstruct had target specificity in both a cell binding assay and cell staining. MicroPET imaging demonstrated VEGFR-specific delivery of the agent to tumors with increasing %ID/g values as a function of time, as well as the expected uptake in the RES (Figure 2.16). Good correlation was observed between PET and NIRF imaging, and data acquired from whole-body imaging were supported by histological examination, which indicated that QD-VEGF primarily targets the tumor vasculature through a VEGF–VEGFR interaction. In both of these studies, the authors showed the feasibility of labeling QDs for PET imaging. They also provided evidence that the attachment of ^{64}Cu-DOTA to the nanoparticles could be readily performed using standard techniques and remain intact under physiologic conditions. Furthermore, they developed vascular-targeted agents using peptide and protein targeting, with both approaches exhibiting target specificity.

A report by Duconge et al. described the use of ^{18}F-based PET imaging of QDs that were encapsulated in a phospholipid micelle formulation to provide highly versatile surface chemistry [149]. The coating strategy resulted in extended circulation time and reduced RES uptake. PET imaging was effective for quantitative whole-body distribution and PK assessment, while the cellular uptake kinetics were determined using in vivo fiber-based confocal fluorescence imaging. The authors suggest that the ability of these QDs to evade opsonization and capture by the RES—as compared to QDs coated with polymeric moieties, which exhibit much faster uptake by the RES—could enhance the efficiency of payload delivery, though the system must be optimized to minimize background signal to achieve better contrast. The continued development of bimodal imaging platforms such as these has substantial diagnostic utility for whole-body PET imaging to assess the extent of disease, followed by cellular or subcellular fluorescence-based imaging to guide surgery and provide histological validation.

Fluorescent silicon QDs are expected to be ideal for many biological applications because of their fluorescent properties and biocompatibility. To evaluate their in vivo fate, Tu and colleagues prepared dextran-coated silicon QDs and labeled them with

(a)

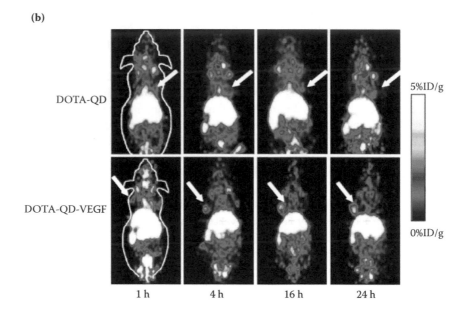

(b)

FIGURE 2.16   (See color insert.) (a) Structure of DOTA-QD-VEGF conjugate. (b) Whole-body coronal PET images of U87MG tumor-bearing mice at 1, 4, 16, and 24 hr after injection of ^{64}Cu-DOTA-QD and ^{64}Cu-DOTA-QD-VEGF. Arrows indicate the tumor. (Adapted with permission from Chen, K. et al. 2008. *European Journal of Nuclear Medicine and Molecular Imaging* 12:2235–2244.)

^{64}Cu for PET imaging [150]. To overcome the limitations of commercially available DOTA analogs, the authors prepared a novel chelating agent that was highly stable and demonstrated no demetalation under ethylenediaminetetraacetic acid (EDTA) challenge. PET imaging revealed that the ^{64}Cu-labeled silicon QDs were rapidly cleared from the blood with high renal filtration and liver uptake, and was in agreement with ex vivo quantification of resected tissues. From these data, the authors concluded that the presence of various species with differences in hydrodynamic diameters, which occurred during agent production, affected the biodistribution profile of the agent, with the fraction of silicon QDs with small hydrodynamic diameter undergoing renal excretion and the fraction with larger hydrodynamic diameter taken up by the RES and liver.

Bimodal QDs for SPECT imaging have also been reported to take advantage of the longer half-life of radionuclides such as ^{125}I for long-term monitoring of in vivo distribution, though low labeling yields and a marked reduction in fluorescence intensity were observed and may limit the effectiveness of this labeling strategy [151]. Nonetheless, there are multiple approaches that can be employed for developing radiolabeled nanoparticles for PET and SPECT imaging and ensuring that they possess a formulation that is physiologically compatible. Improvements in chelator design and conjugation approaches (i.e., click chemistry) are expected to simplify coupling reactions and improve reaction yields, thus enabling preparation of high specific activity compounds that can be administered at lower doses to reduce concerns over toxicity.

2.6 SUMMARY AND FUTURE PERSPECTIVES

Nanoparticles have shown the ability to overcome many of the limitations facing other drug delivery approaches and are becoming increasingly important in medicine. Nuclear imaging has been a vital part of this emergence as it has helped researchers in the nanotechnology field gain a better understanding of in vivo properties that (1) effectively predict utility and (2) guide agent optimization strategies. Given the importance of PET and SPECT imaging in clinical medicine, ongoing efforts in radiopharmaceutical chemistry, molecular imaging agent design, and imaging instrumentation and software development are expected to build upon current capabilities and directly benefit nanomedicine. However, to characterize the clinical potential of promising nanoparticle platforms fully, several key considerations must be taken into account. For example, an advantage of nanoparticles is their ability to be produced in different shapes and sizes, and with various surface modifications. This, however, adds complexity to the agent characterization process as uniformity may not be easily achieved between different batches of nanoparticles that have been functionalized, for example, with multiple chelating agents, targeting moieties, and PEG linkers. In the case of chelators, the number of moieties present will directly impact radiolabeling efficiency and specific activity, which is critical for molecular imaging. Variability in this area will likely impact imaging performance of radiolabeled nanoparticles, thus, accurate chelator quantification must be performed. As chelator conjugation is dependent on the number of reactive sites on a nanoparticle and may vary based on how the nanoparticle was produced, chemical characterization is critical to ensure that the plethora of in vivo data being generated with comparable nanoconjugates can be fairly compared so that conclusions can be accurately drawn.

Another key aspect that must be considered in the translation of radiolabeled nanoparticles is the selection of appropriate in vivo models to characterize the efficacy of drug delivery. In cancer, the majority of nanoparticle imaging agents are nontargeted and utilize the EPR effect to achieve target localization; however, the understanding of EPR effects in different preclinical models (i.e., subcutaneous vs. orthotopic implant) is unclear [152]. Further studies comparing efficacy in different animal models will be instrumental in deciding which approach best resembles clinical cases. Furthermore, since cancer is likely to be the major disease model where the utility of radiolabeled nanoparticles is determined, greater use of tumor models that focus on orthotopic implants and metastatic disease are needed. By mimicking the relevant interactions between stroma and tumor cells, orthotopic models are capable of exhibiting rapid tumor growth and metastasizing to regional and distant lymph nodes to recapitulate clinical scenarios more accurately. This also enables more effective evaluation of in vivo imaging by taking into account the clearance properties of a given nanoparticle and determining its ability to obtain adequate contrast for not only primary tumors, but also metastases throughout the body. This is not the case with xenograft models, which typically place a tumor in the extremities or flank region, where background is inherently low, and organs involved in physiological clearance of the agent are not factored into its ability to delineate the lesion.

Finally, there are significant concerns over the potential side effects caused by nanoparticle toxicity. Toxic effects can arise from the nanoparticle itself, or from breakdown products caused by metabolic processes, and the severity may depend on how fast they are cleared from the body. To that end, the application of surface modifications has helped improve the biocompatibility of many nanoparticle platforms, though large-scale toxicity studies are needed to determine the suitability of a particular approach for human use. Studies that examine the impact of optimized surface chemistry approaches must occur in tandem with nuclear imaging to define the toxicological properties of nanoparticles and realize translational goals. In order to achieve the long-term goal of integrating nanomedicine into routine clinical diagnosis and therapy, multidisciplinary approaches must continue with a keen focus on satisfying the necessary regulatory requirements to provide efficacy and ensure patient safety.

ACKNOWLEDGMENTS

The authors would like to thank Dr. Eva Sevick-Muraca and Dr. Melissa Aldrich for their critical and editorial review of this work.

REFERENCES

1. Phelps, M. E. 2000. PET: The merging of biology and imaging into molecular imaging. *Journal of Nuclear Medicine* 4:661–681.
2. Blankenberg, F. G. and H. W. Strauss. 2007. Nuclear medicine applications in molecular imaging: 2007 update. *Quarterly Journal of Nuclear Medicine and Molecular Imaging* 2:99–110.
3. Imam, S. K. 2005. Molecular nuclear imaging: The radiopharmaceuticals (review). *Cancer Biotherapy and Radiopharmacy* 2:163–172.
4. Delbeke, D., H. Schoder, W. H. Martin, and R. L. Wahl. 2009. Hybrid imaging (SPECT/CT and PET/CT): Improving therapeutic decisions. *Seminars in Nuclear Medicine* 5:308–340.

5. Deleye, S., R. Van Holen, J. Verhaeghe, S. Vandenberghe, S. Stroobants, and S. Staelens. 2013. Performance evaluation of small-animal multipinhole muSPECT scanners for mouse imaging. *European Journal of Nuclear Medicine and Molecular Imaging* 40(5):744–758.
6. Kitson, S. L., V. Cuccurullo, A. Ciarmiello, D. Salvo, and L. Mansi. 2009. Clinical applications of positron emission tomography (PET) imaging in medicine: Oncology, brain diseases and cardiology. *Current Radiopharmacy* 4:224–253.
7. Kato, T., J. Shinoda, N. Nakayama, K. Miwa, A. Okumura, H. Yano, S. Yoshimura, et al. 2008. Metabolic assessment of gliomas using 11C-methionine, [18F] fluorodeoxyglucose, and 11C-choline positron-emission tomography. *American Journal of Neuroradiology* 6:1176–1182.
8. Biersack, H. J. and L. M. Freeman. 2007. *Clinical nuclear medicine*. London: Springer Limited.
9. Quigley, H., S. J. Colloby, and J. T. O'Brien. 2011. PET imaging of brain amyloid in dementia: A review. *International Journal of Geriatric Psychiatry* 10:991–999.
10. Kepe, V., Y. Bordelon, A. Boxer, S.-C. Huang, J. Liu, F.C. Thiede, J.C. Mazziotta, et al. 2013. PET imaging of neuropathology in tauopathies: Progressive supranuclear palsy. *Journal of Alzheimer's Disease* 36 (1):145-153.
11. Dhar, R. and K. Ananthasubramaniam. 2011. Rubidium-82 cardiac positron emission tomography imaging: An overview for the general cardiologist. *Cardiology Review* 5:255–263.
12. Siegrist, P., L. Husmann, M. Knabenhans, O. Gaemperli, I. Valenta, T. Hoefflinghaus, H. Scheffel, et al. 2008. 13N-ammonia myocardial perfusion imaging with a PET/CT scanner: impact on clinical decision making and cost-effectiveness. *European Journal of Nuclear Medicine and Molecular Imaging* 5:889–895.
13. Segall, G. 2002. Assessment of myocardial viability by positron emission tomography. *Nuclear Medicine Communications* 4:323–330.
14. Kubota, K. 2001. From tumor biology to clinical PET: A review of positron emission tomography (PET) in oncology. *Annals of Nuclear Medicine* 6:471–486.
15. Krause, B. J., S. Schwarzenbock, and M. Souvatzoglou. 2013. FDG PET and PET/CT. *Recent Results Cancer Research*:351–369.
16. US Department of Health and Human Services, Food and Drug Administration, Center for Drug Evaluation and Research. December 2009. *Guidance, PET drugs—Current good manufacturing practice (CGMP)*. Available from http://www.fda.gov/downloads/Drugs/GuidanceComplianceRegulatoryInformation/Guidances/ucm070306.pdf.
17. Asit, K. P. and A.-N. Hani. 2007. Cancer imaging agents for positron emission tomography: Beyond FDG. *Current Medical Imaging Reviews* 3:178–185.
18. Tateishi, U., T. Oka, and T. Inoue. 2012. Radiolabeled RGD peptides as integrin alpha(v) beta3-targeted PET tracers. *Current Medicinal Chemistry* 20:3301–3309.
19. Quon, A. and S. S. Gambhir. 2005. FDG-PET and beyond: Molecular breast cancer imaging. *Journal of Clinical Oncology* 8:1664–1673.
20. Mankoff, D. A., J. F. Eary, J. M. Link, M. Muzi, J. G. Rajendran, A. M. Spence, and K. A. Krohn. 2007. Tumor-specific positron emission tomography imaging in patients: [18F] fluorodeoxyglucose and beyond. *Clinical Cancer Research* 12:3460–3469.
21. Gabriel, M., C. Decristoforo, D. Kendler, G. Dobrozemsky, D. Heute, C. Uprimny, P. Kovacs, et al. 2007. 68Ga-DOTA-Tyr3-octreotide PET in neuroendocrine tumors: Comparison with somatostatin receptor scintigraphy and CT. *Journal of Nuclear Medicine* 4:508–518.
22. Anderson, C. J., F. Dehdashti, P. D. Cutler, S. W. Schwarz, R. Laforest, L. A. Bass, J. S. Lewis, and D.W. McCarthy. 2001. 64Cu-TETA-Octreotide as a PET imaging agent for patients with neuroendocrine tumors. *Journal of Nuclear Medicine* 2:213–221.

23. Wu, A. M. 2009. Antibodies and antimatter: The resurgence of immuno-PET. *Journal of Nuclear Medicine* 1:2–5.
24. Levin, C. S. and E. J. Hoffman. 1999. Calculation of positron range and its effect on the fundamental limit of positron emission tomography system spatial resolution. *Physics in Medicine and Biology* 3:781.
25. Anger, H. O. 1958. Scintillation camera. *Review of Scientific Instruments* 1:27–33.
26. Bailey, D. L. and K. P. Willowson. 2013. An evidence-based review of quantitative SPECT imaging and potential clinical applications. *Journal of Nuclear Medicine* 1:83–89.
27. Cherry, S. R., J. Sorenson, and M. Phelps. 2003. *Physics in nuclear medicine,* 3rd ed. Philadelphia: Saunders.
28. Mariani, G., L. Bruselli, T. Kuwert, E. E. Kim, A. Flotats, O. Israel, M. Dondi, and N. Watanabe. 2010. A review on the clinical uses of SPECT/CT. *European Journal of Nuclear Medicine and Molecular Imaging* 10:1959–1985.
29. Mortimer, J. E., F. Dehdashti, B. A. Siegel, K. Trinkaus, J. A. Katzenellenbogen, and M. J. Welch. 2001. Metabolic flare: Indicator of hormone responsiveness in advanced breast cancer. *Journal of Clinical Oncology* 11:2797–2803.
30. Baum, R. P., H. R. Kulkarni, and C. Carreras. 2012. Peptides and receptors in image-guided therapy: Theranostics for neuroendocrine neoplasms. *Seminars in Nuclear Medicine* 3:190–207.
31. Kipriyanov, S. M., S. Dubel, F. Breitling, R. E. Kontermann, and M. Little. 1994. Recombinant single-chain Fv fragments carrying C-terminal cysteine residues: Production of bivalent and biotinylated miniantibodies. *Molecular Immunology* 14:1047–1058.
32. Olafsen, T., C. W. Cheung, P. J. Yazaki, L. Li, G. Sundaresan, S. S. Gambhir, M. A. Sherman, et al. 2004. Covalent disulfide-linked anti-CEA diabody allows site-specific conjugation and radiolabeling for tumor targeting applications. *Protein Engineering Design and Selection* 1:21–27.
33. Olafsen, T. and A. M. Wu. 2010. Antibody vectors for imaging. *Seminars in Nuclear Medicine* 3:167–181.
34. Adair, J. R., P. W. Howard, J. A. Hartley, D. G. Williams, and K. A. Chester. 2012. Antibody-drug conjugates—A perfect synergy. *Expert Opinion in Biological Therapy* 9:1191–1206.
35. Sun, M. M., K. S. Beam, C. G. Cerveny, K. J. Hamblett, R. S. Blackmore, M. Y. Torgov, F. G. Handley, et al. 2005. Reduction-alkylation strategies for the modification of specific monoclonal antibody disulfides. *Bioconjugate Chemistry* 5:1282–1290.
36. Kwekkeboom, D. J., W. W. de Herder, C. H. van Eijck, B. L. Kam, M. van Essen, J. J. Teunissen, and E. P. Krenning. 2010. Peptide receptor radionuclide therapy in patients with gastroenteropancreatic neuroendocrine tumors. *Seminars in Nuclear Medicine* 2:78–88.
37. Ambrosini, V., M. Fani, S. Fanti, F. Forrer, and H. R. Maecke. 2011. Radiopeptide imaging and therapy in Europe. *Journal of Nuclear Medicine* 52:42S–55S.
38. Signore, A., A. Annovazzi, M. Chianelli, F. Corsetti, C. Van de Wiele, and R. N. Watherhouse. 2001. Peptide radiopharmaceuticals for diagnosis and therapy. *European Journal of Nuclear Medicine* 10:1555–1565.
39. Aina, O. H., T. C. Sroka, M. L. Chen, and K. S. Lam. 2002. Therapeutic cancer targeting peptides. *Biopolymers* 3:184–199.
40. Arap, W., R. Pasqualini, and E. Ruoslahti. 1998. Cancer treatment by targeted drug delivery to tumor vasculature in a mouse model. *Science* 5349:377–380.
41. Liu, J., C. Kolar, T. A. Lawson, and W. H. Gmeiner. 2001. Targeted drug delivery to chemoresistant cells: Folic acid derivatization of FdUMP[10] enhances cytotoxicity toward 5-FU-resistant human colorectal tumor cells. *Journal of Organic Chemistry* 17:5655–5663.

42. Kim, B. Y., J. T. Rutka, and W. C. Chan. 2010. Nanomedicine. *New England Journal of Medicine* 25:2434–2443.
43. Wadas, T. J., E. H. Wong, G. R. Weisman, and C. J. Anderson. 2007. Copper chelation chemistry and its role in copper radiopharmaceuticals. *Currents in Pharmaceutical Design* 1:3–16.
44. Psimadas, D., P. Georgoulias, V. Valotassiou, and G. Loudos. 2012. Molecular nanomedicine towards cancer: (1)(1)(1)In-labeled nanoparticles. *Journal of Pharmaceutical Science* 7:2271–2280.
45. Chakraborty, S. and S. Liu. 2010. (99m)Tc and (111)In-labeling of small biomolecules: Bifunctional chelators and related coordination chemistry. *Current Topics in Medicinal Chemistry* 11:1113–1134.
46. Stoker, H. S. 2012. *General, organic, and biological chemistry.* Stamford, CT: Brooks Cole Publishing Company.
47. Banerjee, S., M. R. Pillai, and N. Ramamoorthy. 2001. Evolution of Tc-99m in diagnostic radiopharmaceuticals. *Seminars in Nuclear Medicine* 4:260–277.
48. Plut, E. M. and G. H. Hinkle. 2000. 111In-capromab pendetide: The evolution of prostate specific membrane antigen and the nuclear imaging of its 111In-labelled murine antibody in the evaluation of prostate cancer. *BioDrugs* 6:437–447.
49. Iagaru, A., S. S. Gambhir, and M. L. Goris. 2008. 90Y-Ibritumomab therapy in refractory non-Hodgkin's lymphoma: Observations from 111In-ibritumomab pretreatment imaging. *Journal of Nuclear Medicine* 11:1809–1812.
50. Boswell, C.A. and M.W. Brechbiel. 2005. Auger electrons: Lethal, low energy, and coming soon to a tumor cell nucleus near you. *Journal of Nuclear Medicine* 12:1946–1947.
51. Rogers, B. E., C. J. Anderson, J. M. Connett, L. W. Guo, W. B. Edwards, E. L. Sherman, K. R. Zinn, and M. J. Welch. 1996. Comparison of four bifunctional chelates for radiolabeling monoclonal antibodies with copper radioisotopes: Biodistribution and metabolism. *Bioconjugate Chemistry* 4:511–522.
52. Fani, M., L. Del Pozzo, K. Abiraj, R. Mansi, M. L. Tamma, R. Cescato, B. Waser, et al. 2011. PET of somatostatin receptor-positive tumors using 64Cu- and 68Ga-somatostatin antagonists: The chelate makes the difference. *Journal of Nuclear Medicine* 7:1110–1118.
53. Varagnolo, L., M. P. Stokkel, U. Mazzi, and E. K. Pauwels. 2000. 18F-labeled radiopharmaceuticals for PET in oncology, excluding FDG. *Nuclear Medicine Biology* 2:103–112.
54. Been, L. B., A. J. Suurmeijer, D. C. Cobben, P. L. Jager, H. J. Hoekstra, and P. H. Elsinga. 2004. [18F]FLT-PET in oncology: Current status and opportunities. *European Journal of Nuclear Medicine and Molecular Imaging* 12:1659–1672.
55. Foo, S. S., D. F. Abbott, N. Lawrentschuk, and A. M. Scott. 2004. Functional imaging of intratumoral hypoxia. *Molecular Imaging Biology* 5:291–305.
56. Mankoff, D. A., T. J. Tewson, and J. F. Eary. 1997. Analysis of blood clearance and labeled metabolites for the estrogen receptor tracer [F-18]-16 alpha-fluoroestradiol (FES). *Nuclear Medicine Biology* 4:341–348.
57. Green, L. A., K. Nguyen, B. Berenji, M. Iyer, E. Bauer, J. R. Barrio, M. Namavari, et al. 2004. A tracer kinetic model for 18F-FHBG for quantitating herpes simplex virus type 1 thymidine kinase reporter gene expression in living animals using PET. *Journal of Nuclear Medicine* 9:1560–1570.
58. Tjuvajev, J. G., M. Doubrovin, T. Akhurst, S. Cai, J. Balatoni, M. M. Alauddin, R. Finn, et al. 2002. Comparison of radiolabeled nucleoside probes (FIAU, FHBG, and FHPG) for PET imaging of HSV1-tk gene expression. *Journal of Nuclear Medicine* 8:1072–1083.

59. Burns, H. D., K. Van Laere, S. Sanabria-Bohorquez, T. G. Hamill, G. Bormans, W. S. Eng, R. Gibson, et al. 2007. [18F]MK-9470, a positron emission tomography (PET) tracer for in vivo human PET brain imaging of the cannabinoid-1 receptor. *Proceedings of National Academy of Sciences USA* 23:9800–9805.

60. Cselenyi, Z., M. E. Jonhagen, A. Forsberg, C. Halldin, P. Julin, M. Schou, P. Johnstrom, et al. 2012. Clinical validation of 18F-AZD4694, an amyloid-beta-specific PET radioligand. *Journal of Nuclear Medicine* 3:415–424.

61. Olafsen, T., D. Betting, V. E. Kenanova, F. B. Salazar, P. Clarke, J. Said, A. A. Raubitschek, et al. 2009. Recombinant anti-CD20 antibody fragments for small-animal PET imaging of B-cell lymphomas. *Journal of Nuclear Medicine* 9:1500–1508.

62. Koehler, L., K. Gagnon, S. McQuarrie, and F. Wuest. 2010. Iodine-124: A promising positron emitter for organic PET chemistry. *Molecules* 4:2686–2718.

63. Palestro, C. J. 1994. The current role of gallium imaging in infection. *Seminars in Nuclear Medicine* 2:128–141.

64. Front, D. and O. Israel. 1995. The role of Ga-67 scintigraphy in evaluating the results of therapy of lymphoma patients. *Seminars in Nuclear Medicine* 1:60–71.

65. Chianelli, M., S.J. Mather, J. Martin-Comin, and A. Signore. 1997. Radiopharmaceuticals for the study of inflammatory processes: A review. *Nuclear Medicine Communications* 5:437–455.

66. Zhang, Y., H. Hong, and W. Cai. 2011. PET tracers based on Zirconium-89. *Currents in Radiopharmacology* 2:131–139.

67. Dijkers, E. C., T. H. Oude Munnink, J. G. Kosterink, A. H. Brouwers, P. L. Jager, J. R. de Jong, G. A. van Dongen, et al. 2010. Biodistribution of 89Zr-trastuzumab and PET imaging of HER2-positive lesions in patients with metastatic breast cancer. *Clinical Pharmacology Therapy* 5:586–592.

68. Borjesson, P. K., Y. W. Jauw, R. Boellaard, R. de Bree, E. F. Comans, J. C. Roos, J. A. Castelijns, et al. 2006. Performance of immuno-positron emission tomography with zirconium-89-labeled chimeric monoclonal antibody U36 in the detection of lymph node metastases in head and neck cancer patients. *Clinical Cancer Research* 7 Pt 1:2133–2140.

69. Borjesson, P. K., Y. W. Jauw, R. de Bree, J. C. Roos, J. A. Castelijns, C. R. Leemans, G. A. van Dongen, and R. Boellaard. 2009. Radiation dosimetry of 89Zr-labeled chimeric monoclonal antibody U36 as used for immuno-PET in head and neck cancer patients. *Journal of Nuclear Medicine* 11:1828–1836.

70. Caride, V. J., W. Taylor, J. A. Cramer, and A. Gottschalk. 1976. Evaluation of liposome-entrapped radioactive tracers as scanning agents. Part 1: Organ distribution of liposome (99mTc-DTPA) in mice. *Journal of Nuclear Medicine* 12:1067–1072.

71. McDougall, I. R., J. K. Dunnick, M. L. Goris, and J. P. Kriss. 1975. In vivo distribution of vesicles loaded with radiopharmaceuticals: A study of different routes of administration. *Journal of Nuclear Medicine* 6:488–491.

72. Espinola, L. G., J. Beaucaire, A. Gottschalk, and V. J. Caride. 1979. Radiolabeled liposomes as metabolic and scanning tracers in mice. II. In-111 oxine compared with Tc-99m DTPA, entrapped in multilamellar lipid vesicles. *Journal of Nuclear Medicine* 5:434–440.

73. Richardson, V. J., B. E. Ryman, R. F. Jewkes, K. Jeyasingh, M. N. Tattersall, E. S. Newlands, and S. B. Kaye. 1979. Tissue distribution and tumor localization of 99m-technetium-labeled liposomes in cancer patients. *British Journal of Cancer* 1:35–43.

74. Lopez-Berestein, G., L. Kasi, M. G. Rosenblum, T. Haynie, M. Jahns, H. Glenn, R. Mehta, et al. 1984. Clinical pharmacology of 99mTc-labeled liposomes in patients with cancer. *Cancer Research* 1:375–378.

75. Phillips, W. T. 1999. Delivery of gamma-imaging agents by liposomes. *Advanced Drug Delivery Reviews* 1–3:13–32.

76. Petersen, A. L., A. E. Hansen, A. Gabizon, and T. L. Andresen. 2012. Liposome imaging agents in personalized medicine. *Advanced Drug Delivery Reviews* 13:1417–1435.
77. Rudolph, A. S., R. W. Klipper, B. Goins, and W. T. Phillips. 1991. In vivo biodistribution of a radiolabeled blood substitute: 99mTc-labeled liposome-encapsulated hemoglobin in an anesthetized rabbit. *Proceedings of National Academy Sciences USA* 23:10976–10980.
78. Goins, B., R. Klipper, A. S. Rudolph, and W. T. Phillips. 1994. Use of technetium-99m-liposomes in tumor imaging. *Journal of Nuclear Medicine* 9:1491–1498.
79. Tilcock, C., M. Yap, M. Szucs, and D. Utkhede. 1994. PEG-coated lipid vesicles with encapsulated technetium-99m as blood pool agents for nuclear medicine. *Nuclear Medicine Biology* 2:165–170.
80. Oyen, W. J., O. C. Boerman, G. Storm, L. van Bloois, E. B. Koenders, R. A. Claessens, R. M. Perenboom, et al. 1996. Detecting infection and inflammation with technetium-99m-labeled stealth liposomes. *Journal of Nuclear Medicine* 8:1392–1397.
81. Kleiter, M. M., D. Yu, L. A. Mohammadian, N. Niehaus, I. Spasojevic, L. Sanders, B. L. Viglianti, et al. 2006. A tracer dose of technetium-99m-labeled liposomes can estimate the effect of hyperthermia on intratumoral doxil extravasation. *Clinical Cancer Research* 22:6800–6807.
82. Hnatowich, D. J., B. Friedman, B. Clancy, and M. Novak. 1981. Labeling of preformed liposomes with Ga-67 and Tc-99m by chelation. *Journal of Nuclear Medicine* 9:810–814.
83. Laverman, P., E. T. Dams, W. J. Oyen, G. Storm, E. B. Koenders, R. Prevost, J. W. van der Meer, et al. 1999. A novel method to label liposomes with 99mTc by the hydrazino nicotinyl derivative. *Journal of Nuclear Medicine* 1:192–197.
84. Dams, E. T., M. M. Reijnen, W. J. Oyen, O. C. Boerman, P. Laverman, G. Storm, J. W. van der Meer, et al. 1999. Imaging experimental intraabdominal abscesses with 99mTc-PEG liposomes and 99mTc-HYNIC IgG. *Annals of Surgery* 4:551–557.
85. Laverman, P., E. T. Dams, G. Storm, T. G. Hafmans, H. J. Croes, W. J. Oyen, F. H. Corstens, and O. C. Boerman. 2001. Microscopic localization of PEG-liposomes in a rat model of focal infection. *Journal of Control Release* 3:347–355.
86. Sikkink, C. J., M. M. Reijnen, P. Laverman, W. J. Oyen, and H. van Goor. 2009. Tc-99m-PEG-liposomes target both adhesions and abscesses and their reduction by hyaluronate in rats with fecal peritonitis. *Journal of Surgery Research* 2:246–251.
87. Laverman, P., A. H. Brouwers, E. T. Dams, W. J. Oyen, G. Storm, N. van Rooijen, F. H. Corstens, and O. C. Boerman. 2000. Preclinical and clinical evidence for disappearance of long-circulating characteristics of polyethylene glycol liposomes at low lipid dose. *Journal of Pharmacology and Experimental Therapy* 3:996–1001.
88. Hwang, K. J., J. E. Merriam, P. L. Beaumier, and K. F. Luk. 1982. Encapsulation, with high efficiency, of radioactive metal ions in liposomes. *Biochimica et Biophysica Acta* 1:101–109.
89. Beaumier, P. L. and K. J. Hwang. 1982. An efficient method for loading indium-111 into liposomes using acetylacetone. *Journal of Nuclear Medicine* 9:810–815.
90. Harrington, K. J., G. Rowlinson-Busza, K. N. Syrigos, P. S. Uster, R. G. Vile, and J. S. Stewart. 2000. PEGylated liposomes have potential as vehicles for intratumoral and subcutaneous drug delivery. *Clinical Cancer Research* 6:2528–2537.
91. Proffitt, R. T., L. E. Williams, C. A. Presant, G. W. Tin, J. A. Uliana, R. C. Gamble, and J. D. Baldeschwieler. 1983. Tumor-imaging potential of liposomes loaded with In-111-NTA: Biodistribution in mice. *Journal of Nuclear Medicine* 1:45–51.
92. Presant, C. A., R. T. Proffitt, A. F. Turner, L. E. Williams, D. Winsor, J. L. Werner, P. Kennedy, et al. 1988. Successful imaging of human cancer with indium-111-labeled phospholipid vesicles. *Cancer* 5:905–911.

93. Kubo, A., K. Nakamura, T. Sammiya, M. Katayama, T. Hashimoto, S. Hashimoto, H. Kobayashi, and T. Teramoto. 1993. Indium-111-labelled liposomes: Dosimetry and tumour detection in patients with cancer. *European Journal of Nuclear Medicine* 2:107–113.

94. Presant, C. A., D. Blayney, R. T. Proffitt, A. F. Turner, L. E. Williams, H. I. Nadel, P. Kennedy, et al. 1990. Preliminary report: Imaging of Kaposi sarcoma and lymphoma in AIDS with indium-111-labelled liposomes. *Lancet* 8701:1307–1309.

95. Khalifa, A., D. Dodds, R. Rampling, J. Paterson, and T. Murray. 1997. Liposomal distribution in malignant glioma: Possibilities for therapy. *Nuclear Medicine Communications* 1:17–23.

96. Jensen, G. M. and T. H. Bunch. 2007. Conventional liposome performance and evaluation: Lessons from the development of Vescan. *Journal of Liposome Research* 3–4:121–137.

97. Helbok, A., C. Decristoforo, G. Dobrozemsky, C. Rangger, E. Diederen, B. Stark, R. Prassl, and E. von Guggenberg. 2010. Radiolabeling of lipid-based nanoparticles for diagnostics and therapeutic applications: A comparison using different radiometals. *Journal of Liposome Research* 3:219–227.

98. Mitchell, N., T. L. Kalber, M. S. Cooper, K. Sunassee, S. L. Chalker, K. P. Shaw, K. L. Ordidge, et al. 2013. Incorporation of paramagnetic, fluorescent and PET/SPECT contrast agents into liposomes for multimodal imaging. *Biomaterials* 4:1179–1192.

99. Seo, J. W., H. Zhang, D. L. Kukis, C. F. Meares, and K. W. Ferrara. 2008. A novel method to label preformed liposomes with 64Cu for positron emission tomography (PET) imaging. *Bioconjugate Chemistry* 12:2577–2584.

100. Rygh, C. B., S. Qin, J. W. Seo, L. M. Mahakian, H. Zhang, R. Adamson, J. Q. Chen, et al. 2011. Longitudinal investigation of permeability and distribution of macromolecules in mouse malignant transformation using PET. *Clinical Cancer Research* 3:550–559.

101. Oku, N., Y. Tokudome, H. Tsukada, T. Kosugi, Y. Namba, and S. Okada. 1996. In vivo trafficking of long-circulating liposomes in tumor-bearing mice determined by positron emission tomography. *Biopharmaceutics & Drug Disposition* 5:435–441.

102. Oku, N., Y. Tokudome, H. Tsukada, and S. Okada. 1995. Real-time analysis of liposomal trafficking in tumor-bearing mice by use of positron emission tomography. *Biochimica et Biophysica Acta* 1:86–90.

103. Marik, J., M. S. Tartis, H. Zhang, J. Y. Fung, A. Kheirolomoom, J. L. Sutcliffe, and K. W. Ferrara. 2007. Long-circulating liposomes radiolabeled with [18F]fluorodipalmitin ([18F]FDP). *Nuclear Medicine Biology* 2:165–171.

104. Urakami, T., S. Akai, Y. Katayama, N. Harada, H. Tsukada, and N. Oku. 2007. Novel amphiphilic probes for [18F]-radiolabeling preformed liposomes and determination of liposomal trafficking by positron emission tomography. *Journal of Medicinal Chemistry* 26:6454–6457.

105. Elbayoumi, T. A. and V. P. Torchilin. 2009. Tumor-specific anti-nucleosome antibody improves therapeutic efficacy of doxorubicin-loaded long-circulating liposomes against primary and metastatic tumor in mice. *Molecular Pharmacology* 1:246–254.

106. Iyer, A. K., Y. Su, J. Feng, X. Lan, X. Zhu, Y. Liu, D. Gao, et al. 2011. The effect of internalizing human single chain antibody fragment on liposome targeting to epithelioid and sarcomatoid mesothelioma. *Biomaterials* 10:2605–2613.

107. Petersen, A. L., T. Binderup, R. I. Jolck, P. Rasmussen, J. R. Henriksen, A. K. Pfeifer, A. Kjaer, and T. L. Andresen. 2012. Positron emission tomography evaluation of somatostatin receptor targeted 64Cu-TATE-liposomes in a human neuroendocrine carcinoma mouse model. *Journal of Control Release* 2:254–263.

108. Helbok, A., C. Rangger, E. von Guggenberg, M. Saba-Lepek, T. Radolf, G. Thurner, F. Andreae, et al. 2012. Targeting properties of peptide-modified radiolabeled liposomal nanoparticles. *Nanomedicine* 1:112–118.

109. de Barros, A. B., A. Tsourkas, B. Saboury, V. N. Cardoso, and A. Alavi. 2012. Emerging role of radiolabeled nanoparticles as an effective diagnostic technique. *EJNMMI Research* 1:39.

110. Bianco, A., K. Kostarelos, C. D. Partidos, and M. Prato. 2005. Biomedical applications of functionalised carbon nanotubes. *Chemistry Communications (Camb)* 5:571–577.

111. Wang, H., J. Wang, X. Deng, H. Sun, Z. Shi, Z. Gu, Y. Liu, and Y. Zhao. 2004. Biodistribution of carbon single-wall carbon nanotubes in mice. *Journal of Nanoscience and Nanotechnology* 8:1019–1024.

112. Mackeyev, Y. A., J. W. Marks, M. G. Rosenblum, and L. J. Wilson. 2005. Stable containment of radionuclides on the nanoscale by cut single-wall carbon nanotubes. *Journal of Physical Chemistry B* 12:5482–5484.

113. Singh, R., D. Pantarotto, L. Lacerda, G. Pastorin, C. Klumpp, M. Prato, A. Bianco, and K. Kostarelos. 2006. Tissue biodistribution and blood clearance rates of intravenously administered carbon nanotube radiotracers. *Proceedings of National Academy Sciences USA* 9:3357–3362.

114. McDevitt, M. R., D. Chattopadhyay, J. S. Jaggi, R. D. Finn, P. B. Zanzonico, C. Villa, D. Rey, et al. 2007. PET imaging of soluble yttrium-86-labeled carbon nanotubes in mice. *PLoS One* 9:e907.

115. Hong, S. Y., G. Tobias, K. T. Al-Jamal, B. Ballesteros, H. Ali-Boucetta, S. Lozano-Perez, P. D. Nellist, et al. 2010. Filled and glycosylated carbon nanotubes for in vivo radioemitter localization and imaging. *Nature Materials* 6:485–490.

116. Liu, Z., W. Cai, L. He, N. Nakayama, K. Chen, X. Sun, X. Chen, and H. Dai. 2007. In vivo biodistribution and highly efficient tumour targeting of carbon nanotubes in mice. *Nature Nanotechnology* 1:47–52.

117. Ruggiero, A., C. H. Villa, J. P. Holland, S. R. Sprinkle, C. May, J. S. Lewis, D. A. Scheinberg, and M. R. McDevitt. 2010. Imaging and treating tumor vasculature with targeted radiolabeled carbon nanotubes. *International Journal of Nanomedicine* 5:783–802.

118. McDevitt, M. R., D. Chattopadhyay, B. J. Kappel, J. S. Jaggi, S. R. Schiffman, C. Antczak, J. T. Njardarson, et al. 2007. Tumor targeting with antibody-functionalized, radiolabeled carbon nanotubes. *Journal of Nuclear Medicine* 7:1180–1189.

119. Guo, J., X. Zhang, Q. Li, and W. Li. 2007. Biodistribution of functionalized multiwall carbon nanotubes in mice. *Nuclear Medicine Biology* 5:579–583.

120. Wu, W., R. Li, X. Bian, Z. Zhu, D. Ding, X. Li, Z. et al. 2009. Covalently combining carbon nanotubes with anticancer agent: Preparation and antitumor activity. *ACS Nano* 9:2740–2750.

121. Al-Jamal, K. T., A. Nunes, L. Methven, H. Ali-Boucetta, S. Li, F. M. Toma, M. A. Herrero, et al. 2012. Degree of chemical functionalization of carbon nanotubes determines tissue distribution and excretion profile. *Angewandte Chemie* International Edition England 26:6389–6393.

122. Tomalia, D. A. 2005. Birth of a new macromolecular architecture: Dendrimers as quantized building blocks for nanoscale synthetic polymer chemistry. *Progress in Polymer Science* 3–4:294–324.

123. Roberts, J. C., Y. E. Adams, D. Tomalia, J. A. Mercer-Smith, and D. K. Lavallee. 1990. Using starburst dendrimers as linker molecules to radiolabel antibodies. *Bioconjugate Chemistry* 5:305–308.

124. Biricova, V., A. Laznickova, M. Laznicek, M. Polasek, and P. Hermann. 2011. Radiolabeling of PAMAM dendrimers conjugated to a pyridine-*N*-oxide DOTA analog with (1)(1)(1)In: Optimization of reaction conditions and biodistribution. *Journal of Pharmaceutical and Biomedical Analysis* 3:505–512.

125. Kobayashi, H., N. Sato, T. Saga, Y. Nakamoto, T. Ishimori, S. Toyama, K. Togashi, et al. 2000. Monoclonal antibody-dendrimer conjugates enable radiolabeling of antibody with markedly high specific activity with minimal loss of immunoreactivity. *European Journal of Nuclear Medicine* 9:1334–1339.

126. Criscione, J. M., L. W. Dobrucki, Z. W. Zhuang, X. Papademetris, M. Simons, A. J. Sinusas, and T. M. Fahmy. 2011. Development and application of a multimodal contrast agent for SPECT/CT hybrid imaging. *Bioconjugate Chemistry* 9:1784–1792.

127. Okuda, T., S. Kawakami, N. Akimoto, T. Niidome, F. Yamashita, and M. Hashida. 2006. PEGylated lysine dendrimers for tumor-selective targeting after intravenous injection in tumor-bearing mice. *Journal of Control Release* 3:330–336.

128. Agashe, H. B., A. K. Babbar, S. Jain, R. K. Sharma, A. K. Mishra, A. Asthana, M. Garg, et al. 2007. Investigations on biodistribution of technetium-99m-labeled carbohydrate-coated poly(propylene imine) dendrimers. *Nanomedicine* 2:120–127.

129. Dijkgraaf, I., A. Y. Rijnders, A. Soede, A. C. Dechesne, G. W. van Esse, A. J. Brouwer, F. H. Corstens, et al. 2007. Synthesis of DOTA-conjugated multivalent cyclic-RGD peptide dendrimers via 1,3-dipolar cycloaddition and their biological evaluation: Implications for tumor targeting and tumor imaging purposes. *Organic and Biomolecular Chemistry* 6:935–944.

130. Zhang, Y., Y. Sun, X. Xu, X. Zhang, H. Zhu, L. Huang, Y. Qi, and Y. M. Shen. 2010. Synthesis, biodistribution, and microsingle photon emission computed tomography (SPECT) imaging study of technetium-99m labeled PEGylated dendrimer poly(amidoamine) (PAMAM)-folic acid conjugates. *Journal of Medicinal Chemistry* 8:3262–3272.

131. Amirkhanov, N. V., K. Zhang, M. R. Aruva, M. L. Thakur, and E. Wickstrom. 2010. Imaging human pancreatic cancer xenografts by targeting mutant KRAS2 mRNA with [(111)In]DOTA(n)-poly(diamidopropanoyl)(m)-KRAS2   PNA-D(Cys-Ser-Lys-Cys) nanoparticles. *Bioconjugate Chemistry* 4:731–740.

132. Gupta, A. K. and M. Gupta. 2005. Synthesis and surface engineering of iron oxide nanoparticles for biomedical applications. *Biomaterials* 18:3995–4021.

133. Stark, D. D., R. Weissleder, G. Elizondo, P. F. Hahn, S. Saini, L. E. Todd, J. Wittenberg, and J. T. Ferrucci. 1988. Superparamagnetic iron oxide: Clinical application as a contrast agent for MR imaging of the liver. *Radiology* 2:297–301.

134. Weissleder, R., P. F. Hahn, D. D. Stark, G. Elizondo, S. Saini, L. E. Todd, J. Wittenberg, and J. T. Ferrucci. 1988. Superparamagnetic iron oxide: Enhanced detection of focal splenic tumors with MR imaging. *Radiology* 2:399–403.

135. Colombo, M., S. Carregal-Romero, M. F. Casula, L. Gutierrez, M. P. Morales, I. B. Bohm, J. T. Heverhagen, et al. 2012. Biological applications of magnetic nanoparticles. *Chemical Society Reviews* 11:4306–4334.

136. Glaus, C., R. Rossin, M. J. Welch, and G. Bao. 2010. In vivo evaluation of (64)Cu-labeled magnetic nanoparticles as a dual-modality PET/MR imaging agent. *Bioconjugate Chemistry* 4:715–722.

137. Lee, H. Y., Z. Li, K. Chen, A. R. Hsu, C. Xu, J. Xie, S. Sun, and X. Chen. 2008. PET/MRI dual-modality tumor imaging using arginine-glycine-aspartic (RGD)-conjugated radio-labeled iron oxide nanoparticles. *Journal of Nuclear Medicine* 8:1371–1379.

138. Yang, X., H. Hong, J. J. Grailer, I. J. Rowland, A. Javadi, S. A. Hurley, Y. Xiao, et al. 2011. cRGD-functionalized, DOX-conjugated, and (6)(4)Cu-labeled superparamagnetic iron oxide nanoparticles for targeted anticancer drug delivery and PET/MR imaging. *Biomaterials* 17:4151–4160.

139. Yang, M., K. Cheng, S. Qi, H. Liu, Y. Jiang, H. Jiang, J. Li, et al. 2013. Affibody modified and radiolabeled gold–Iron oxide hetero-nanostructures for tumor PET, optical and MR imaging. *Biomaterials* 11:2796–2806.

140. Torres Martin de Rosales, R., R. Tavare, A. Glaria, G. Varma, A. Protti, and P. J. Blower. 2011. ((9)(9)m)Tc-bisphosphonate-iron oxide nanoparticle conjugates for dual-modality biomedical imaging. *Bioconjugate Chemistry* 3:455–465.

141. Sandiford, L., A. Phinikaridou, A. Protti, L. K. Meszaros, X. Cui, Y. Yan, G. Frodsham, et al. 2013. Bisphosphonate-anchored PEGylation and radiolabeling of superparamagnetic iron oxide: Long-circulating nanoparticles for in vivo multimodal (T1 MRI-SPECT) imaging. *ACS Nano* 1:500–512.

142. Madru, R., P. Kjellman, F. Olsson, K. Wingardh, C. Ingvar, F. Stahlberg, J. Olsrud, et al. 2012. 99mTc-labeled superparamagnetic iron oxide nanoparticles for multimodality SPECT/MRI of sentinel lymph nodes. *Journal of Nuclear Medicine* 3:459–463.

143. Alivisatos, A. P., K. P. Johnsson, X. Peng, T. E. Wilson, C. J. Loweth, M. P. Bruchez, Jr., and P. G. Schultz. 1996. Organization of "nanocrystal molecules" using DNA. *Nature* 6592:609–611.

144. Ballou, B., B. C. Lagerholm, L. A. Ernst, M. P. Bruchez, and A. S. Waggoner. 2004. Noninvasive imaging of quantum dots in mice. *Bioconjugate Chemistry* 1:79–86.

145. Allen, P. M., W. Liu, V. P. Chauhan, J. Lee, A. Y. Ting, D. Fukumura, R. K. Jain, and M. G. Bawendi. 2010. InAs(ZnCdS) quantum dots optimized for biological imaging in the near-infrared. *Journal of American Chemical Society* 2:470–471.

146. Bentolila, L. A., Y. Ebenstein, and S. Weiss. 2009. Quantum dots for in vivo small-animal imaging. *Journal of Nuclear Medicine* 4:493–496.

147. Cai, W., K. Chen, Z. B. Li, S. S. Gambhir, and X. Chen. 2007. Dual-function probe for PET and near-infrared fluorescence imaging of tumor vasculature. *Journal of Nuclear Medicine* 11:1862–1870.

148. Chen, K., Z. B. Li, H. Wang, W. Cai, and X. Chen. 2008. Dual-modality optical and positron emission tomography imaging of vascular endothelial growth factor receptor on tumor vasculature using quantum dots. *European Journal of Nuclear Medicine and Molecular Imaging* 12:2235–2244.

149. Duconge, F., T. Pons, C. Pestourie, L. Herin, B. Theze, K. Gombert, B. Mahler, et al. 2008. Fluorine-18-labeled phospholipid quantum dot micelles for in vivo multimodal imaging from whole body to cellular scales. *Bioconjugate Chemistry* 9:1921–1926.

150. Tu, C., X. Ma, A. House, S. M. Kauzlarich, and A. Y. Louie. 2011. PET imaging and biodistribution of silicon quantum dots in mice. *ACS Medicinal Chemistry Letters* 4:285–288.

151. Park, J. J., T. S. Lee, J. H. Kang, R. Song, and G. J. Cheon. 2011. Radioiodination and biodistribution of quantum dots using Bolton–Hunter reagent. *Applied Radiative Isotopes* 1:56–62.

152. Prabhakar, U., H. Maeda, R. K. Jain, E. M. Sevick-Muraca, W. Zamboni, O. C. Farokhzad, S. T. Barry, et al. 2013. Challenges and key considerations of the enhanced permeability and retention effect for nanomedicine drug delivery in oncology. *Cancer Research* 8:2412–2417.

3 Molecular Imaging at Nanoscale with Magnetic Resonance Imaging

Patrick M. Winter

CONTENTS

3.1 INTRODUCTION

The aim of magnetic resonance (MR) molecular imaging is to noninvasively map the cellular expression of important biomarkers associated with normal physiology or disease. Standard MR imaging methods lack the resolution and sensitivity required for direct detection of these cellular biomarkers due to their very low concentrations in vivo. To overcome this obstacle, MR molecular imaging often relies on specifically engineered nanoparticles that can (1) bind to the biomarker of interest, (2) accumulate at the target site, and (3) generate sufficient image contrast. Nanoparticle constructs allow for the incorporation of multiple binding ligands, which can improve the targeting ability, and multiple relaxation agents, which amplify the MR signal enhancement. MR molecular imaging contrast agents generally utilize either paramagnetic gadolinium chelates or superparamagnetic iron oxide to generate image contrast. Most iron oxide contrast agents consist of rigid structures that are outside the scope of this chapter, but have been extensively described in the literature [1–5]. Gadolinium agents have been grafted onto a wide range of nanoparticle constructs, including liposomes, micelles, dendrimers, polymers, viral particles, and liquid perfluorocarbon particles.

In some cases, the nanoparticle may also serve as a drug delivery agent, providing diagnostic and therapeutic information via noninvasive MR imaging. Selective targeting of a therapeutic drug could improve uptake at the pathological tissue and reduce retention in normal tissues. These properties could combine to reduce the required dose and/or lower the occurrence of unwanted side effects for highly potent drugs. The ability to image the uptake of drug-laden nanoparticles in the body could be utilized to directly evaluate the biodistribution and elimination of the agent in tissues of interest and to predict the end-organ therapeutic effect.

3.2   PARAMAGNETIC PERFLUOROCARBON NANOPARTICLES

Perfluorocarbon (PFC) nanoparticles contain a liquid PFC core encapsulated within a monolayer of phospholipids. These nanoparticles have been utilized for a diverse array of biomedical applications, including as a targeted molecular imaging contrast agent and tissue-specific drug delivery vehicle. The size of PFC particles (250 nm in diameter) typically limits their biodistribution to the intravascular space. Therefore, these agents are often formulated with targeting molecules that specifically bind intraluminal biomarkers, such as integrins, selectins, or adhesion molecules [6,7]. Chemically, PFC molecules are similar to hydrocarbons, but with fluorine atoms in place of the hydrogens [6–8]. PFCs are highly stable and biologically inert compounds due to the dense electron cloud of carbon–fluorine bonds, which impede chemical reactions that could break down the molecular structure [6].

PFC nanoparticles have been specifically formulated for molecular imaging of a variety of disease biomarkers and for use with many different imaging modalities [6,7]. In particular, this contrast agent has been utilized for imaging angiogenesis, a hallmark of many disease processes, including, but not limited to, tumor growth and atherosclerosis [6]. To achieve molecular imaging of angiogenesis, PFC nanoparticles can be formulated with a targeting agent that selectively binds to the $\alpha_v\beta_3$-integrin. Expression of $\alpha_v\beta_3$-integrins on the vascular lumen is required for the migration and invasion of vascular endothelial cells, a vital step in the formation of new blood vessels [9]. Targeted PFC nanoparticle contrast agents have been designed to be detected by ultrasound imaging [10], MR [7], computed tomography [11], nuclear imaging [12], and optical imaging [13–15] as well as for multimodality imaging.

In order to detect PFC nanoparticles with MR, a lipophilic gadolinium chelate is typically added to the outer phospholipid monolayer. Gadolinium is a paramagnetic metal that induces increased T1 relaxation of the tissue, leading to a hyperintense signal on T1-weighted MR images. The ability of paramagnetic agents to accelerate T1 relaxation of the MR signal is denoted by the relaxivity of the agent, measured in units of per second per millimolar, $1/(s*mM)$. Anchoring the gadolinium onto a nanoparticle carrier can improve the relaxivity of the contrast agent via two mechanisms. First, the relatively large size of the particle slows the tumbling rate of the metal, improving the chemical interaction between the gadolinium ion and the surrounding water molecules. As a result, the relaxivity of Gd-DTPA at 1.5 T increases from 4.5 $1/(s*mM)$ when diluted in water [16] to 17.9 $1/(s*mM)$ when incorporated onto PFC nanoparticles [17]. Second, utilizing a nanoparticle allows multiple copies of the gadolinium chelate to be loaded onto a single carrier, effectively multiplying the paramagnetic effect induced in the target tissue each time a particle binds to the biomarker of interest. PFC nanoparticles can be loaded with nearly 100,000 Gd-DTPA molecules per particle, leading to an overall relaxivity of 1,690,000 $1/(s*mM)$ at 1.5T [17]. With such a high relaxivity, the minimum concentration of

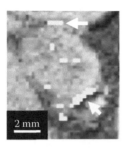

FIGURE 3.1 T1-weighted MR image of rabbit tumor 2 hours after injection of $\alpha_v\beta_3$-integrin-targeted PFC nanoparticles. The yellow overlay indicates the areas of MR signal enhancement, which were typically observed along the tumor periphery near blood vessels or along borders with surrounding musculature (white arrows). (Reprinted with permission from Winter, P. M. et al. 2003. *Cancer Research* 63 (18): 5838–5843.)

PFC nanoparticles required for MR image detection, defined as the concentration required to produce a contrast-to-noise ratio of 5, is around 100 pM at a typical clinical field strength (1.5 T) and 25 pM at a higher field strength used for high-resolution imaging in animal research (4.7 T) [17].

Paramagnetic PFC nanoparticles targeted to $\alpha_v\beta_3$-integrins have been used for molecular imaging of tumor angiogenesis in rabbits using a clinical 1.5 T MR scanner [18]. T1-weighted MR images of animals receiving $\alpha_v\beta_3$-integrin-targeted paramagnetic nanoparticles displayed a marked increase in the MR signal in a patchy distribution mostly along the tumor periphery (Figure 3.1). Two hours after nanoparticle injection, the MR signal increased by 126% relative to the preinjection image intensity. Histology of the tumors revealed that angiogenesis was sparsely distributed along the tumor periphery and often occurred near large blood vessels that were outside the tumor capsule (Figure 3.2). Nontargeted paramagnetic nanoparticles yielded significantly lower MR signal enhancement, demonstrating the specific binding of the agent to the $\alpha_v\beta_3$-integrin. Furthermore, in vivo competition studies demonstrated that binding of the targeted particles could be blocked by pretreatment with $\alpha_v\beta_3$-integrin-targeted nonparamagnetic nanoparticles. The MR signal intensity of the adjacent muscle tissue remained constant following injection of either $\alpha_v\beta_3$-integrin-targeted or nontargeted nanoparticles, indicating that this agent does not accumulate in normal tissues. This study demonstrated that $\alpha_v\beta_3$-integrin-targeted PFC nanoparticles could provide high target-specific avidity with robust signal enhancement on a clinical imaging system for the sensitive detection of tumor angiogenesis in vivo.

Research studies on molecular imaging with targeted nanoparticles have also explored various applications in predicting, monitoring, or improving a range of therapeutic methods. For example, molecular imaging of tumor angiogenesis could provide noninvasive monitoring of the efficacy of antiangiogenic drugs, such as avastin, sutent, or nexavar. Early detection of the therapeutic response would allow personalized tailoring of drug selection and/or dosing to maximize the antitumor effect and possibly reduce the drug side effects. Furthermore, the targeted contrast agent could serve as a carrier platform for delivery of the drug itself. Antiangiogenic therapy has been demonstrated in tumor-bearing rabbits using $\alpha_v\beta_3$-integrin-targeted

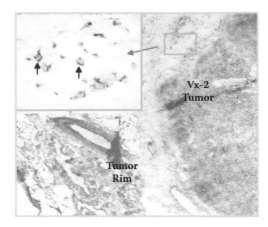

FIGURE 3.2 **(See color insert.)** Histology of rabbit tumor showing angiogenesis (black arrows) near a large blood vessel adjacent to the tumor rim. Tumor sections were stained with H&E (low magnification image) to show morphology or LM-609 (inset, high magnification image) to show $\alpha_v\beta_3$-integrin expression. The anatomical location of angiogenic vasculature determined via histology corresponds to the areas of MR signal enhancement following injection of $\alpha_v\beta_3$-integrin-targeted nanoparticles. (Reprinted with permission from Winter, P. M. et al. 2003. *Cancer Research* 63 (18): 5838–5843.)

nanoparticles that carried the drug fumagillin in the outer phospholipid membrane of the particles [19]. Fumagillin binds to methionine aminopeptidase 2 (MetAP2), a key enzyme required for the growth of new blood vessels, and inhibits angiogenesis. Treatment with $\alpha_v\beta_3$-integrin-targeted fumagillin nanoparticles reduced the tumor volume by 52%–66% compared to animals receiving nontargeted fumagillin nanoparticles, $\alpha_v\beta_3$-integrin-targeted nanoparticles without drug or saline. Due to the targeted delivery of fumagillin, this antitumor effect was achieved with a drug dose that was 1,000 times lower than in other animal studies and 60 times lower than in clinical trials [19]. MR molecular imaging with paramagnetic $\alpha_v\beta_3$-integrin-targeted nanoparticles allowed in vivo mapping of the antiangiogenic effect of the fumagillin treatment (Figure 3.3). Fumagillin tumors displayed sparse angiogenesis along the tumor periphery, while control tumors showed a dense distribution of angiogenesis covering an area 2.5 times larger than in treated tumors. Fluorescence microscopy confirmed that the $\alpha_v\beta_3$-integrin-targeted nanoparticles were preferentially retained in the tumor periphery (Figure 3.4), as observed by MR. Furthermore, the nanoparticles were colocalized with FITC-lectin, demonstrating that this agent is confined to the intravascular space.

Similarly to the neovascular needs of growing tumors, atherosclerotic plaques require extensive angiogenesis to meet the increased metabolic demands of the developing lesions. In atherosclerosis, angiogenesis is typically observed in the vasa vasorum of large arteries, including the aorta, carotids, and coronaries. MR molecular imaging of angiogenesis in the vasa vasorum of an atherosclerotic animal model has been reported using $\alpha_v\beta_3$-integrin-targeted PFC nanoparticles [20]. The targeted particles generated a 47% increase in the MR signal in the aortic wall,

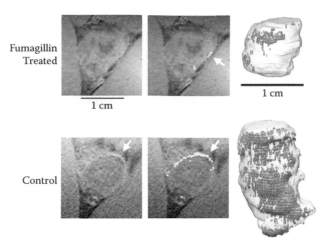

FIGURE 3.3   (See color insert.) Reduced MR contrast enhancement in T1-weighted images of a rabbit treated with $\alpha_v\beta_3$-integrin-targeted fumagillin nanoparticles (top) compared to an animal receiving $\alpha_v\beta_3$-integrin-targeted nanoparticles without drug (bottom). Enhancing pixels, color coded in yellow (white arrows), demonstrate large areas of angiogenesis in the control tumor and markedly lower levels of angiogenesis with fumagillin treatment. Panels on the right-hand side demonstrate 3D neovascular maps of the tumors with and without fumagillin therapy. The angiogenic pixels are color coded in blue and are much more prevalent in the periphery of the control tumor than the treated tumor. (Reprinted with permission from Winter, P. M. et al. 2008. *FASEB Journal* 22 (8): 2758–2767.)

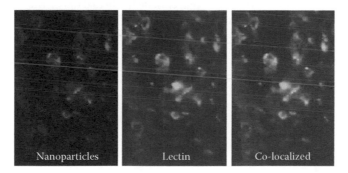

FIGURE 3.4   (See color insert.) Fluorescence microscopy (20 times magnification) of the tumor periphery showing $\alpha_v\beta_3$-integrin-targeted nanoparticles containing rhodamine (left) and FITC-lectin (middle). Overlaying the fluorescent signal from these two agents demonstrates that $\alpha_v\beta_3$-integrin-targeted nanoparticles are constrained to the vasculature and taken up by angiogenic capillaries. (Reprinted with permission from Winter, P. M. et al. 2008. *FASEB Journal* 22 (8): 2758–2767.)

while nontargeted particles and competitive blockade of the $\alpha_v\beta_3$-integrin-binding sites resulted in much lower signal enhancement. In a follow-up study, PFC nanoparticles were formulated with a fluorescent dye to follow particle binding by microscopic examination of tissue sections [14]. As expected for an agent of this size, the nanoparticles were constrained to the vascular space and did not extravasate into

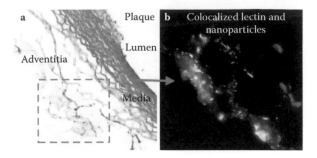

FIGURE 3.5 Histology of atherosclerotic aorta following injection with $\alpha_v\beta_3$-integrin-targeted PFC nanoparticles containing rhodamine- and fluorescein isothiocyanate (FITC)-labeled lectin, a marker of vascular endothelial cells. Nanoparticle binding was localization of nanoparticle binding with respect to normal and pathological vascular structures. (a) Light microscopy with hematoxylin and eosin (H&E) staining shows the vessel microstructure, including luminal plaque, media, and adventitia. (b) The fluorescent image demonstrates colocalization of the rhodamine-labeled $\alpha_v\beta_3$-integrin-targeted nanoparticles with FITC-labeled lectin. (Reprinted with permission from Winter, P. M. et al. 2008. *JACC Cardiovascular Imaging* 1 (5): 624–634.)

the media or intima of the vessel wall. The fluorescent signal from the nanoparticles colocalized with FITC-lectin, a vascular marker, in the adventitia demonstrating neovascular proliferation in the vasa vasorum (Figure 3.5). Histology revealed a strong correlation between the MR signal enhancement and the density of angiogenic microvessels in the vasa vasorum (Figure 3.6) [14]. The MR signal steadily increased at higher angiogenic densities, but dramatically decreased at lower densities.

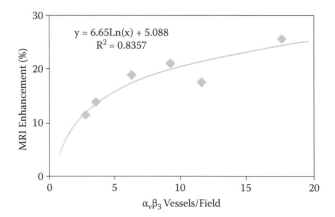

FIGURE 3.6 MR molecular imaging and histology revealed a logarithmic relationship between the MR signal enhancement and the microvascular density. Aortic sections were stained for $\alpha_v\beta_3$-integrin and platelet endothelial cell adhesion molecule (PECAM), to quantify the angiogenic microvasculature. (Reprinted with permission from Winter, P. M. et al. 2008. *JACC Cardiovascular Imaging* 1 (5): 624–634.)

MR molecular imaging of angiogenesis could be used to map serially the abundance of biomarkers associated with atherosclerosis and/or therapeutic response. Thus, the biochemical signatures of disease could be monitored noninvasively to provide early and highly specific detection of pathology, rather than tracking patient symptoms or the occurrence of clinical events. Molecular imaging of angiogenesis has been used to monitor the physiological effects of a range of therapeutic treatments for atherosclerosis, including an appetite suppressant, L-arginine, atorvastatin, and antiangiogenic fumagillin therapy [7].

Paramagnetic $\alpha_v\beta_3$-integrin-targeted PFC nanoparticles have been used to map angiogenesis in the aortic wall of atherosclerotic rats and serially monitor the effects of benfluorex, an appetite suppressant [21]. The JCR:LA-cp rat is a model of metabolic syndrome that displays obesity, insulin resistance, hyperlipidemia, and vasculopathy. Obese rats are homozygous for a defective leptin receptor and display increased food consumption, body weight, insulin, leptin, cholesterol, and triglycerides compared to control rats that have at least one copy of the normal gene. MR molecular imaging with $\alpha_v\beta_3$-integrin-targeted nanoparticles revealed widespread angiogenesis in the aortic wall of the obese animals, indicating actively developing atherosclerotic plaques (Figure 3.7). Benfluorex reduced food intake, body weight, insulin, and leptin in the obese treated rats, but did not change the serum cholesterol or triglyceride levels. Likewise, benfluorex reduced aortic angiogenesis in the obese treated animals to the same level as in the control lean rats. Histological analysis of the microvascular density in the aortic wall confirmed the MR molecular imaging

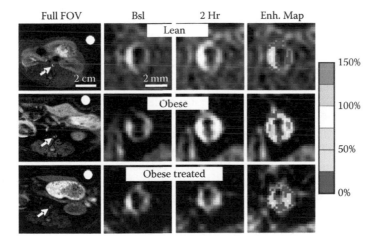

FIGURE 3.7 T1-weighted, fat-suppressed MR image (full FOV) shows large abdominal fat deposits in obese (middle row) and obese treated (bottom row) JCR rats that were not observed in the lean (top row) animals. The arrows denote the abdominal aorta in each example. Magnified views of the aortas are shown at baseline (Bsl) and 2-hr postinjection of $\alpha_v\beta_3$-integrin-targeted nanoparticles, demonstrating MR signal enhancement in the aorta wall. Quantitation of the signal enhancement in percentage (Enh. Map; overlaid onto postinjection image) reveals markedly higher MR signal enhancement in the obese animal, due to active angiogenesis in response to the development of atherosclerotic plaques. (Reprinted with permission from Cai, K. et al. 2010. *JACC Cardiovascular Imaging* 3 (8): 824–832.)

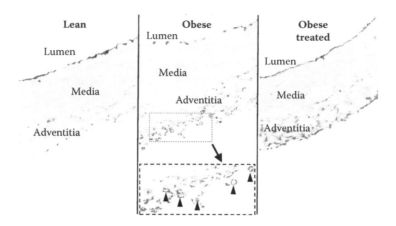

FIGURE 3.8 Histological staining of endothelial cells in the aortic wall of lean (left), obese (middle), and obese treated (right) JCR rats corroborates the MR molecular imaging results. The obese animal displayed prominent angiogenesis in the adventitia (arrowheads) that was not observed in the lean phenotype. (Reprinted with permission from Cai, K. et al. 2010. *JACC Cardiovascular Imaging* 3 (8): 824–832.)

results (Figure 3.8), corroborating that $\alpha_v\beta_3$-integrin-targeted nanoparticles can be used to monitor atherosclerosis and the neovascular effects of an appetite suppression therapy noninvasively.

However, the biological consequences of atherosclerosis are not always associated with higher levels of angiogenesis in tissues. In a rabbit model of peripheral vascular disease, atherosclerosis leads to reduced angiogenesis in the ischemic limb. This abnormal response can be normalized with a proangiogenic therapy, such as L-arginine supplied via the drinking water [22]. MR molecular imaging with paramagnetic $\alpha_v\beta_3$-integrin-targeted PFC nanoparticles revealed significantly higher angiogenesis in the ischemic limb of L-arginine-treated rabbits compared to controls (Figure 3.9). Furthermore, the integrin-targeted formulation produced two times higher MR enhancement compared to nontargeted particles, demonstrating the advantage of biomarker targeting. X-ray angiography showed that L-arginine animals developed more collateral vessels compared to the untreated controls. Histologic staining of muscle capillaries revealed a denser pattern of microvasculature in L-arginine-treated animals (Figure 3.10), confirming the MR and x-ray imaging results. Noninvasive molecular imaging with PFC nanoparticles could be used for early detection of response to angiogenic therapies directed at ischemic diseases, such as peripheral vascular disease or myocardial infarction, to enable personalized optimization of treatment strategies.

In addition to the anticancer therapy described previously, PFC nanoparticles formulated with fumagillin have been studied as a targeted therapeutic in atherosclerotic rabbits. Since antiangiogenic therapies have been successfully employed to slow the progression of cancer, it is reasonable to investigate if these treatments could also reverse the pathological effects of atherosclerosis. The acute antiangiogenic effect of $\alpha_v\beta_3$-integrin-targeted paramagnetic fumagillin nanoparticles was

Molecular imaging of angiogenesis

Right leg   Left leg   Right leg   Left leg
(Ligated)   (Control)   (Ligated)   (Control)

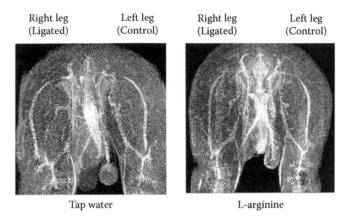

Tap water   L-arginine

FIGURE 3.9   (See color insert.) MR molecular imaging enhancement (color coded in red) shows diffuse angiogenesis throughout the ischemic leg (right) with only slight enhancement in the control leg (left). The animal receiving tap water (left panel) shows more enhancement in the ligated leg compared to the control leg, but L-arginine treatment (right panel) results in a more dense distribution of angiogenesis in the ligated limb, while the control limb appears similar to the untreated animal. (Reprinted with permission from Winter, P. M. et al. 2010. *Magnetic Resonance Medicine* 64 (2): 369–376.)

H&E staining   CD31 staining
Tap water   Tap water   L-arginine

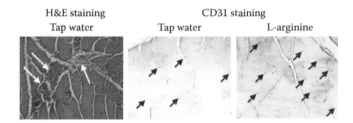

FIGURE 3.10   (See color insert.) Histology of muscle from the ischemic limb of animals treated with tap water or L-arginine. Left: Hematoxylin and eosin (H&E) staining shows intramuscular hemorrhage (white arrows) in tap water animals, which was not observed with L-arginine treatment. Middle and right: Microvascular staining (black arrows) revealed more capillaries in L-arginine-treated animals compared to animals receiving tap water, supporting the MR molecular imaging and x-ray angiography findings that L-arginine augments angiogenic response to limb ischemia. (Reprinted with permission from Winter, P. M. et al. 2010. *Magnetic Resonance Medicine* 64 (2): 369–376.)

evaluated by examining expansion of the vasa vasorum in the aortic wall [23]. MR images collected at the time of targeted fumagillin treatment were used to estimate drug deposition in the aortic wall. One week later, after treatment, the expression of the $\alpha_v\beta_3$-integrin was measured via MR molecular imaging with $\alpha_v\beta_3$-integrin-targeted paramagnetic nanoparticles (Figure 3.11). MR signal enhancement in the fumagillin-treated rabbits decreased by 83% compared to pretreatment value. In

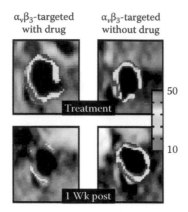

FIGURE 3.11 MR molecular imaging of $\alpha_v\beta_3$-integrin expression in the abdominal aorta of atherosclerotic rabbits before (top) and 1 week after (bottom) treatment with $\alpha_v\beta_3$-integrin-targeted fumagillin nanoparticles (left) or $\alpha_v\beta_3$-integrin-targeted nanoparticles without drug (right). The intensity and distribution of the MR signal enhancement is shown as a false-colored overlay of percent signal enhancement. (Reprinted with permission from Winter, P. M. et al. 2006. *Arteriosclerosis, Thrombosis and Vascular Biology* 26 (9): 2103–2109.)

comparison, rabbits receiving no treatment showed an 8% increase in enhancement and rabbits receiving nontargeted fumagillin nanoparticles showed a 26% decrease in the MR signal. The dramatic reduction of $\alpha_v\beta_3$-integrin expression in the aortic wall was confirmed by histology, which revealed that the neovascular density was 67% lower in the aorta of treated rabbits compared to nontreated animals. Furthermore, aortic areas that produced the highest MR enhancement at the time of treatment subsequently displayed the largest reduction in $\alpha_v\beta_3$-integrin-targeted MR signal 1 week later (Figure 3.12). These findings suggest that a combined molecular imaging and targeted therapy agent can be used to confirm and quantify the local delivery of chemotherapeutics as well as provide early predictions of the end organ treatment response.

In addition to evaluating the acute response to therapy, MR molecular imaging has been used for serial monitoring of the antiangiogenic effects of atorvastatin and targeted fumagillin nanoparticles [14]. The effect of continuous atorvastatin treatment was studied with and without fumagillin treatment performed once every 4 weeks. During the 8-week study, atorvastatin did not reduce $\alpha_v\beta_3$-integrin expression in the aortic wall. Fumagillin treatment, however, caused lower image enhancement that persisted for 2–3 weeks, demonstrating a successful although shortlived antiangiogenic effect (Figure 3.13). By treating the rabbits with both targeted fumagillin nanoparticles and atorvastatin, a sustained decrease in the MR signal was achieved. These results suggest that chronic statin treatment could prolong the effects of an antiangiogenic therapy that is delivered in discrete doses. The clinical application of an effective and sustained antiangiogenic treatment could decrease intramural hemorrhage and inflammation, leading to regression or stabilization of atherosclerotic plaques.

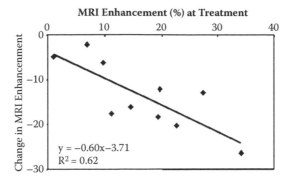

FIGURE 3.12 MR molecular imaging of $\alpha_v\beta_3$-integrin-targeted fumagillin nanoparticles provides confirmation of drug delivery to the aortic wall and prediction of the therapeutic effect observed 1 week after treatment. Regions of the abdominal aorta that displayed the highest signal enhancement at the time of targeted fumagillin treatment also showed the most pronounced reduction of $\alpha_v\beta_3$-integrin expression at follow-up. (Reprinted with permission from Winter, P. M. et al. 2006. *Arteriosclerosis, Thrombosis and Vascular Biology* 26 (9): 2103–2109.)

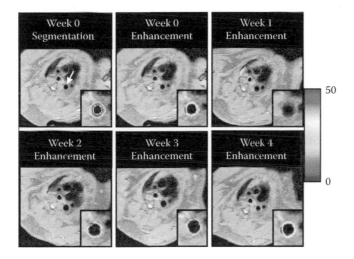

FIGURE 3.13 **(See color insert.)** Serial MR molecular imaging of angiogenesis in the aortic wall before and after targeted fumagillin treatment. In the baseline image (week 0 panels), cross-sectional imaging of the thoracic aorta (arrow) provides robust segmentation of the aortic wall (yellow outline) and reveals patchy areas of angiogenesis (color-coded overlay). Following treatment (week 1 panel), the signal enhancement is markedly lower due to the antiangiogenic effect of fumagillin. The level of signal enhancement gradually increases (week 2 and week 3 panels) and eventually returns to the pretreatment value (week 4 panel). (Reprinted with permission from Cai, K. et al. 2010. *JACC Cardiovascular Imaging* 3 (8): 824–832.)

A wide variety of other applications for targeted PFC nanoparticles have also been explored, including different targeting moieties, other disease models, and alternate MR detection approaches. In addition to $\alpha_v\beta_3$-integrin targeting, paramagnetic PFC nanoparticles have been targeted to the $\alpha_5\beta_1$-integrin [13], VCAM-1 [24], and Robo4 [25]. Furthermore, MR molecular imaging with targeted nanoparticles has been demonstrated in nude mice bearing human melanomas [6,13], atherosclerotic ApoE$^{-/-}$ mice [24], and a mouse model of squamous cell cancer [26]. While MR molecular imaging usually relies on observing the relaxation effect of paramagnetic or superparamagnetic metals on the water signal, PFC nanoparticles can also be detected directly via ^{19}F MR imaging or spectroscopy. The PFC core of the nanoparticles can provide a definitive and quantitative MR signature [7,8]. ^{19}F MR has been utilized to map fibrin deposition on human carotid endarterectomy samples at 4.7 T [27], to differentiate different PFC species targeted to fibrin clots [28], to measure VCAM-1 expression in the mouse kidney at 11.7 T [24], to separate the signals from nanoparticles bound to tissue versus blood pool signal [26], to quantify angiogenesis in the aortic valve of atherosclerotic rabbits [29], and to measure oxygen tension in the blood pool [30]. The flexibility of PFC nanoparticles allows these agents to target various cellular receptors, carry a range of therapeutic agents, and exploit numerous imaging strategies. In the long term, these properties will continue to push the development of PFC nanoparticles for improving the diagnosis and treatment of the most debilitating diseases facing society, including atherosclerosis, cancer, obesity, and diabetes.

3.3 LIPOSOMES

Liposomes have a long and important history in biological research and medicine. They were first described in 1964 [31] and have been utilized for a number of drug delivery [32] and molecular imaging [33] applications. In one of the first demonstrations of MR molecular imaging, a liposome agent containing a paramagnetic chelate and an antibody that binds to the $\alpha_v\beta_3$-integrin was used to map tumor angiogenesis in a rabbit model [34]. More recently, paramagnetic liposomes were targeted to CD-105 and used to image tumor angiogenesis in rats [35]. CD-105 is also known as endoglin and is a cellular glycoprotein required for endothelial cell proliferation [36]. Antibodies targeted against CD-105 are routinely used in histology to assess microvascular density in tissue sections. The CD-105-targeted paramagnetic liposomes produced a progressive increase in the MR signal up to 120 minutes postinjection [35]. The MR enhancement was evident in discrete tumor regions that had histologically verified CD-105 positive cells. Liposomes lacking the CD-105 binding antibody displayed much different MR contrast characteristics, with a diffuse enhancement pattern that peaked at 60 minutes postinjection and then washed out of the tumor. In another application, paramagnetic liposomes were formulated with phosphatidylserine to target atherosclerotic plaques [37]. Phosphatidylserine is expressed on the surface of apoptotic cells, triggering macrophages to engulf the cells. By incorporating phosphatidylserine on the liposomes, the contrast agent would be taken up by macrophages within the plaque or by macrophages in circulation and then carried to the plaque. The abdominal aorta of ApoE$^{-/-}$ mice showed clear MR signal enhancement up to 8 hours postinjection of the liposomes. To confirm the targeting ability of this contrast agent, ex vivo microscopy was used to colocalize the signal from rhodamine included in the liposome phospholipids and a macrophage biomarker, CD-68.

The versatility of liposome constructs has allowed them to be modified for MR detection utilizing contrast materials other than paramagnetic gadolinium. Superparamagnetic iron oxide particles have been extensively studied for molecular imaging [4,38–41] and cell tracking [42–45], but they will generally not be included in this review. However, iron oxide agents have been incorporated inside liposomes for molecular imaging of experimental tumors. For example, liposomes containing saposin C (SapC) in the phospholipid membrane have been developed to bind to cancer cells, such as neuroblastoma cells. SapC is a multifunctional protein known to induce membrane fusion [46]. SapC preferentially interacts with unsaturated, negatively charged phospholipids, such as phosphatidylserine [46,47], which are expressed by tumor cells and the supporting vasculature [48,49]. This affinity for phosphatidylserine allows SapC vesicles to target tumors selectively, but not normal tissues [50,51]. Iron oxide particles were coupled to the liposome phospholipids by oxidizing the dextran coating [52]. The liposomes were then formed with a 200 nm polycarbonate filter and iron particles on the exterior of the liposomes were removed by lowering the pH and filtering the free iron particles via a chromatography column. These superparamagnetic liposomes had a T2 relaxivity of 101 1/(mM*s) at 7 T [50]. In cultured neuroblastoma cells, the T2* relaxation rate increased by 4.9 1/s for each picogram of iron per cell. In tumor-bearing mice, gradient echo images showed a decrease in signal after injection, with the maximum contrast change observed 3 hours postinjection. Quantitation of the in vivo relaxation time showed that the tumor T2 decreased by 30% after injection, 91 ms preinjection versus 62 ms postinjection.

In addition to imaging agents, liposomes have been utilized in a number of drug delivery applications. These particles can be formulated to carry hydrophilic drugs, such as doxorubicin or cytarabine, in the aqueous core [53,54]. Alternatively, hydrophobic drugs, such as temoporfin or flavopiridol, can be incorporated into the outer phospholipid lipid membrane [55,56]. One particularly exciting application is the development of thermosensitive liposomes to control drug release based on thermal activation within the tissue of interest, such as tumors. Liposomes formulated with a phospholipid membrane containing DPPC, HSPC, cholesterol, and DPPE-PEG2000 were used to encapsulate a clinical MR contrast agent, Gd-HPDO3A, and an antitumor chemotherapy agent, doxorubicin [57]. The final concentration of gadolinium and doxorubicin in the liposome solution was 14.2 and 3.1 mM, respectively, indicating successful encapsulation within the particle core. The liposomes displayed a rather weak T1 relaxivity, 0.5 1/(mM*s), at temperatures ranging from 30°C to 39°C, but the relaxivity quickly rose to 3 1/(mM*s) as the temperature increased between 40°C and 43°C. Further temperature increases, up to 50°C, did not produce any additional changes in the relaxivity, indicating that the liposome membrane rapidly becomes very porous at 40°C–43°C, facilitating the release of the gadolinium and doxorubicin payloads. To investigate the in vivo utility of this therapeutic formulation, liposomes were injected into rats bearing subcutaneous 9L tumors and the animals were subjected to MR guided high-intensity focused ultrasound (HIFU) hyperthermia or no HIFU (control group). The HIFU-treated animals displayed varying amounts of increased tumor relaxation rates, ranging from 0.09 to 0.58 1/s. On the other hand, the control animals all showed minimal changes in relaxation rates, averaging 0.03 1/s. Ex vivo tumor analysis showed varying amounts of gadolinium and doxorubicin in the HIFU-treated tissue, 0.37% ID/g to 1.74% ID/g and 0.36% ID/g to 1.91% ID/g, respectively, but much lower concentrations in the control animals, 0.21% ID/g and 0.19% ID/g, respectively. Interestingly, the non-invasive measurement of MR relaxation rates at the time of treatment was highly correlated to the subsequent concentrations of gadolinium ($R^2 = 0.9805$) and doxorubicin ($R^2 = 0.9996$) in the tumors. While these experiments may not have achieved consistent release of the drug and contrast agent payloads within the tumor, they clearly show that the MR signal can provide a surrogate measurement for local drug deposition within the tissue of interest.

In a similar set of experiments, liposomes formulated with DPPC lipid and Brij78 were engineered to release doxorubicin and Gd-DTPA rapidly within a tumor upon HIFU activation [58]. In phantom studies, the release of Gd-DTPA from the particle core was correlated to the doxorubicin release kinetics. Total regression of in vivo tumors was achieved when 5–10 mg doxorubicin per kilogram of bodyweight was released locally in the tumor from the liposome carrier. This drug dose was associated with a 0.03 1/s decrease in the MR relaxation rate of the tumor at the time of treatment. The development of MR contrast agents that quantifiably map the local deposition of drugs in real time could help make personalized therapy an effective and robust tool in the ongoing fight against cancer and other diseases.

3.4   OTHER NANOPARTICLE SYSTEMS

Nature has harnessed a number of nanoparticle structures to accomplish various biological processes, including viruses, lipoproteins, and micelles. In addition, synthetic nanoparticles have been developed for a host of chemical and industrial applications, such as dendrimers and polymers. High-density lipoproteins (HDL) and low-density lipoproteins (LDL) have been labeled with gadolinium for MR imaging of atherosclerosis and cancer. HDL particles consist of a triglyceride and cholesterol core encapsulated by a phospholipid membrane. These particles are utilized to transport lipids from different body tissues to the liver for metabolism and/or excretion. The phospholipid membrane of HDL offers an ideal surface for the incorporation of amphiphilic gadolinium chelates. The apolipoproteins associated with HDL, such as apo A-I and apo A-II, selectively target these particles to atherosclerotic plaques [59]. Paramagnetic HDL particles were injected into atherosclerotic mice, yielding a 35% enhancement in the MR signal intensity of the abdominal aorta. Histology showed that the particles were taken up by macrophages, which make up a large portion of vascular plaques and are associated with plaque rupture. In a similar manner, LDL particles modified to contain amphiphilic gadolinium contrast agents have been developed for imaging tumors. A number of different tumor types, such as prostate, melanoma, colon, and breast, overexpress LDL receptors, making this agent an ideal candidate for selective imaging of these cancers. Mice bearing subcutaneous liver carcinomas were imaged before and 24 hours after injection of paramagnetic LDL particles [60]. The MR signal from the tumor increased by 25% compared to pre-injection. Microscopic analysis of treated cells showed that the particles were internalized within the cytosol, indicating active uptake via the LDL receptor.

Similarly to HDL and LDL particles, micelles consist of a hydrophobic core encapsulated within a phospholipid layer. Surface modification has been exploited to label micelles with paramagnetic chelates and targeting ligands for specific MR imaging of tumors. A paramagnetic micelle agent was targeted to the $\alpha_v\beta_3$-integrin via incorporation of an RGD (arginylglycylaspartic acid) peptide on the surface [61]. The nanoparticles displayed a relaxivity of 12 1/(mM*s) relative to the gadolinium concentration. However, each micelle carried approximately 150 gadolinium ions, yielding a relaxivity around 2000 1/(mM*s) in terms of micelle concentration. The micelles were injected into melanoma-bearing mice and produced significant MR signal enhancement localized mainly along the periphery of the tumor. This agent was formulated also to carry an optical imaging agent, and ex vivo tissue imaging showed that targeted micelles accumulated in tumor blood vessels, but not in muscle. While viruses are generally considered to be harmful and dangerous agents, they can be modified in the lab to eliminate their infectious properties. The remaining nanostructure is highly stable and naturally biocompatible. A number of viral species have been modified for MR imaging, including cowpea chlorotic mottle virus [62,63], MS2 bacteriophage [64], and adenovirus [65].

Dendrimers and polymers have similar chemical formulas, but their structures are quite distinct. Polymers consist of a linear chain of repeated subunits. On the other hand, dendrimers are highly branched structures that take on a spherical conformation. One of the first angiogenic biomarkers ever discovered was vascular endothelial growth factor (VEGF) [66], which plays a prominent role in a variety of diseases, including cancer, ischemia, infection, atherosclerosis, diabetes, inflammation, and organ rejection. Selective targeting of VEGF with MR contrast agents has been accomplished by several techniques [67]. For example, gadolinium-loaded polymeric nanoparticles containing an anti-VEGF targeting ligand have been used to image angiogenesis associated with hepatocellular carcinomas in a mouse model [68]. Human epidermal growth factor receptor type 2 (HER2/neu) is a cell surface receptor that is overexpressed in many types of cancer, including breast, ovarian, bladder, and others. Furthermore, tumors with higher HER2/neu expression levels tend to have a poorer prognosis than tumors with lower expression. Several studies have explored MR molecular imaging of this biomarker by directly labeling the HER2/neu antibody [69] or by using a nanoparticle scaffold [70]. For the nanoparticle approach, a fourth-generation dendrimer was labeled with gadolinium chelates. Of the 64 reactive sites on each dendrimer molecule, the gadolinium bound to 21 sites on average. The relaxivity of the agent was 137 1/(mM*s) relative to the dendrimer concentration. Using a HER2/neu antibody to target the dendrimer to breast tumors in mice yielded a decrease in the tumor T1 from 2.3 seconds to 1.4 seconds 24 hours after injection.

3.5   CLINICAL MOLECULAR IMAGING

Clinical molecular imaging applications have mostly centered on tagging antibodies or other targeting molecules with radioactive elements. For example, single-photon emission computed tomography (SPECT) and positron emission tomography (PET) imaging of various compounds, including glucose, choline, thymidine, and cellular receptors, has been used for identifying cancer, cardiovascular disease, and other disorders [71–77]. Due to the relative insensitivity of MR imaging, labeling a single molecule with a paramagnetic tag typically does not generate sufficient image contrast in clinical applications. There are, however, a couple of specific examples where the biomarker of interest is expressed at such a high level that gadolinium tagging can be employed successfully.

Ablavar is an MR contrast agent approved for use in MR angiography. While Ablavar does not bind to a cellular receptor like the molecular imaging agents discussed previously, it is designed to bind albumin in the blood pool. By binding to this highly abundant serum protein, Ablavar has a long circulatory residence time [78], allowing increased signal-to-noise and/or spatial resolution compared to traditional contrast agents with faster elimination rates. In addition, binding to albumin increases the relaxivity of Ablavar by a factor of five compared to other clinical contrast agents [79]. As a blood pool contrast agent, Ablavar has widespread applications in clinical MR imaging of arterial stenosis, vascular malformations, deep vein thrombosis, and other abnormalities of the blood vessels [80]. While Ablavar does not fit the traditional mold of molecular imaging contrast agents, its approval by regulatory agencies and its subsequent adoption into clinical practice serve as a model for the development and acceptance of other targeted MR molecular imaging agents.

Another contrast agent that takes advantage of a highly abundant biomarker binds to the fibrin that makes up a thrombus. An 11-amino acid peptide chain that binds to human fibrin has been coupled to four gadolinium chelates to form a targeted MR contrast agent called EP-2104R [81]. EP-2104R utilizes a macrocyclic chelator to reduce the potential for gadolinium to be released in the body. Under extreme conditions, including very low pH and a very strong competing chelator, EP-2104R showed very low gadolinium loss—0.3%—compared to other clinically approved contrast agents, 71.7% for Gd-DTPA, 88.4% for Gd-DTPA-BMA, and 39.2% for Gd-DO3A. EP-2104R has strong binding to human fibrin with a K_d of 1.6 μM and cross reactivity with pig, dog, rabbit, rat, and mouse fibrin. In comparison, the binding of EP-2104R to fibrinogen is much weaker, with a K_d of 240 μM. The relaxivity of EP-2104R is 11.1 $1/(mM^*s)$ in solution, but increases to 24.9 $1/(mM^*s)$ when bound to a fibin clot [82,83]. A number of studies in animal models of cardiovascular disease demonstrate the wide variety of potential applications for this agent in clinical imaging. Preclinical studies have shown that EP-2104R could reliably detect pulmonary emboli, coronary thrombi, and thrombi in the right ventricle in pigs at 1.5 T 2 hours after injection [84]. After injection of EP-2104R, the contrast-to-noise ratio of the clots was between 17 and 56 and the gadolinium content was 144 µmol/L compared to 22 µmol/L after injection of Gd-DTPA. In a preclinical model of thrombosis in the cerebral vein of a pig, injection of EP-2104R yielded a contrast-to-noise ratio of 14 [85]. Furthermore, early clinical trials have demonstrated that this agent is easily tolerated in patients with intravascular thrombi. A small cohort of 10 patients with clots in the heart, thoracic aorta, or carotid artery were imaged before and 2–6 hours after injection of EP-2104R [82]. The contrast-to-noise ratio of the thrombi increased from 6 ± 8 before contrast injection to 29 ± 14 after injection. In a study assessing the feasibility for imaging both arterial and venous clots, early postinjection time points—2 to 6 hours—showed a 90% increase in the MR signal intensity of the clots, while later time points—20–36 hours—yielded an increase of 104% [83]. There is a critical medical need to image clots in patients suffering from heart attacks and strokes. A fibrin-specific MR contrast agent could be invaluable for identifying culprit thrombi and for monitoring the effectiveness of "clot-busting" therapies in this large and growing patient population.

3.6 PARACEST AGENTS

One significant limitation of all paramagnetic and superparamagnetic MR molecular imaging contrast agents is that biomarker detection requires comparison of the MR signal intensity before and after injection of the agent. The postinjection image may be collected hours or days after the preinjection image, making image registration and signal comparisons difficult. A new class of MR contrast agent offers the unique opportunity to turn the image contrast on and off by changing the imaging parameters, thus eliminating the need for pre- and postinjection imaging.

Chemical exchange saturation transfer (CEST) agents have exchangeable protons, such as NH or OH groups, that resonate at a chemical shift that is distinguishable from the bulk water signal. However, CEST agents often have chemical shifts that are very close to the bulk water signal, making it difficult to saturate just the exchangeable proton signal without directly impacting the bulk water peak. Coupling a paramagnetic ion near the water-binding sites can shift the bound water frequency further away from the bulk water, allowing selective saturation of only the exchangeable protons [86]. Applying radiofrequency prepulses at the appropriate frequency and power level can saturate the exchangeable protons and transfer the magnetization into the bulk water pool, leading to a reduced MR signal [87]. Therefore, these paramagnetic CEST (PARACEST) agents allow the image contrast to be switched on and off at will. In addition to typical molecular imaging applications, the water exchange kinetics of PARACEST agents can be designed to be sensitive to specific environmental conditions, such as pH [88,89] or temperature [89,90], or to interact with specific metabolites, such as lactate [91], glucose [92], or zinc [93], yielding an MR signal enhancement that is proportional to these biologically important variables.

Liposomes have been employed to entrap PARACEST agents within the hydrophilic core. The permeability of the liposome membrane can be modified by incorporating different phospholipids. With adequate mixing of the interior and exterior water compartments, liposomes can produce a measureable PARACEST signal. These agents are known collectively as lipoCEST agents. The water inside the liposome acts as the "bound" water pool, which resonates at an MR frequency offset from the bulk water signal. Selective saturation of the water molecules inside the particles is transferred to the outside pool through the liposome membrane. For example, a lipoCEST formulation that encapsulated 200 mM of Tm-DOTMA within the particle core generated a PARACEST signal that was 3.1 ppm away from the bulk water [94]. Nonspherical lipoCEST particles produce a bound water peak that is further away from the bulk water signal due to the larger bulk magnetic susceptibility shift of these particles [95]. A nonspherical lipoCEST agent containing Tm-HPDO3A or Dy-HPDO3A displayed a bound water resonance at 30 or −40 ppm, respectively [96]. Other efforts to improve the sensitivity of lipoCEST particles have focused on incorporating amphiphilic shift reagents into the phospholipid membrane [97] or encapsulating polymeric shift reagents within the aqueous core [98]. A lipoCEST agent targeted to angiogenesis was used to image brain tumors in mice [99]. The agent displayed a bound water peak at 7 ppm and produced an MR enhancement of 1.5% within 1–2 hours postinjection. For nontargeted lipoCEST particles or in contralateral brain tissue, the MR enhancement only reached 0.5%–0.75%. LipoCEST agents have shown great flexibility in fine-tuning the water exchange rate and bound water frequency, which could be exploited to develop different formulations for specific targets or drug delivery applications.

One of the unique properties of PARACEST MR contrast agents is that multiple agents can be separately detected based on the offset frequency of the bound water signals. Conventional paramagnetic and superparamagnetic MR agents influence the water relaxation rates, and the effects of individual agents cannot be distinguished from one another. However, two or more PARACEST agents with different bound water frequencies can be individually interrogated with selective saturation pulses, yielding pulse sequences that are sensitive to one contrast agent and not the others. Selective imaging of two dendrimer PARACEST agents has been performed in tumor-bearing mice [100]. A europium-based PARACEST chelate was coupled onto a fifth-generation polyamidoamine (PAMAM) dendrimer, forming Eu-PAMAM. In a similar fashion, an ytterbium chelate was coupled to a second-generation PAMAM dendrimer to form Yb-PAMAM. Eu-PAMAM displayed a bound water peak at 55 ppm, while Yb-PAMAM yielded a bound water peak at −16 ppm. Intravenous injection of a mixture of Eu-PAMAM and Yb-PAMAM showed different pharmacokinetics curves for the bound water signals at 55 ppm and −16 ppm. The smaller Yb-PAMAM dendrimer showed rapid uptake into the tumor, while the larger Eu-PAMAM agent displayed reduced tissue permeability, as expected. Yb-PAMAM also appeared to reach a plateau by the end of the 30-minute MR imaging exam, while the Eu-PAMAM signal continued to increase. Similar methods could be utilized to evaluate the biodistribution of targeted versus nontargeted agents [101,102], to measure tissue pH based on the different PARACEST signals from pH-responsive and -insensitive agents [103,104], or to map metabolic activity utilizing activatable and unalterable chelates [105–107].

Adenovirus particles have been widely utilized in research labs to target specific cell populations and to deliver gene therapy vectors. Viral particles have been modified to carry PARACEST chelates on the outer surface of the capsid and the bioactivity was evaluated in HT 1080 cells [65]. MR imaging of the PARACEST virus particles was performed with an on-resonance WALTZ-16 pulse sequence. The PARACEST virus showed significantly higher contrast than unmodified particles. These experiments demonstrate the feasibility of labeling adenovirus particles with PARACEST chelates while maintaining adequate bioactivity for use in cultured cells.

In cardiovascular disease, MR molecular imaging with PFC nanoparticles targeted to fibrin allows sensitive identification of clots [108,109]. While paramagnetic PFC nanoparticles can produce signal enhancement on T1-weighted MR images, an alternate approach could utilize PARACEST chelates incorporated on the surface of PFC nanoparticles. Fibrin-targeted PARACEST nanoparticles could provide a means to identify clots without the need for comparing the MR signal intensity of pre- and postinjection images. PARACEST nanoparticles formulated with a europium-based chelate yielded a bound water peak at 52 ppm [101]. Human clots were treated with fibrin-targeted PARACEST nanoparticles or control nanoparticles and imaged at 4.7 T. MR images collected with presaturation at –52 ppm appeared very similar, regardless of the nanoparticle treatment (Figure 3.14). The surface of clots treated with fibrin-targeted PARACEST nanoparticles showed clear MR enhancement after subtraction of images collected with the contrast turned off (presaturation at –52 ppm) from images with the contrast turned on (presaturation at 52 ppm). In contradistinction, the surface of clots treated with control nanoparticles could not

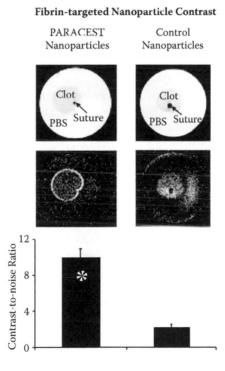

FIGURE 3.14   Molecular imaging of fibrin clots with antibody-targeted PFC PARACEST (left) or control (right) nanoparticles. MR images with presaturation at –52 ppm (top) show no differences between the two clots. However, subtraction images (middle) clearly display signal enhancement on the clot surface treated with PARACEST nanoparticles, but no enhancement on the clot treated with control nanoparticles. The contrast to noise of the clot surface (bottom) was significantly higher with PARACEST nanoparticles compared to control nanoparticles (*P < 0.05). (Reprinted with permission from Winter, P. M. et al. 2006. *Magnetic Resonance Medicine* 56 (6): 1384–1388.)

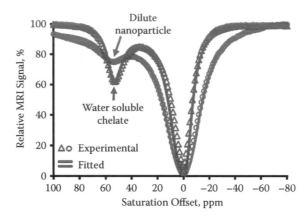

FIGURE 3.15 PARACEST presaturation curve for a water-soluble PARACEST chelate and PFC PARACEST nanoparticles showing a bound water peak at +51 ppm. The experimental data were fit to the Bloch equations to calculate the bound water lifetime of the water-soluble chelate (290 ms) and the PFC nanoparticles (108 ms). (Reprinted with permission from Cai, K. et al. 2011. *NMR Biomedicine* 25:279–285.)

be discriminated from noise upon subtraction. The average PARACEST contrast-to-noise ratio at the clot surface was significantly higher for fibrin-targeted PARACEST nanoparticles (10.0 ± 1.0) compared to the control nanoparticles (2.2 ± 0.4; P < 0.05).

Further experiments revealed that the water exchange kinetics are different for the water-soluble PARACEST chelate, the PARACEST nanoparticle in solution, and the nanoparticles after binding to a biological target [102]. MR signal enhancement was measured for both the water-soluble chelate and PARACEST nanoparticles at 11.7 T over a range of offset frequencies: −80 to 100 ppm (Figure 3.15). Both agents showed a clear bound water peak at 51 ppm. Fitting the experimental data to the modified Bloch equations revealed that the bound water lifetime for the water-soluble chelate was 290 µs, while the nanoparticle agent had a bound water lifetime of 108 µs. Simulation of the Bloch equations indicated that the optimum bound water lifetime was 976 µs—much higher than either of the PARACEST agents studied (Figure 3.16).

For assessing the PARACEST nanoparticles when bound to a clot, the ^{19}F MR signal from the PFC core provided a means to quantitate the nanoparticle concentration at the target site [27,110]. Clots treated with fibrin-targeted nanoparticles showed PARACEST contrast only along the clot boundary—not the clot interior or the surrounding saline (Figure 3.17), generating a PARACEST contrast-to-noise ratio of 17.7 along the clot surface. The ^{19}F images also showed signal only at the clot surface, yielding a signal-to-noise ratio of 7.34 for the targeted clots. The ^{19}F signal indicated that the concentration of particles per voxel at the clot surface was 8.13 n*M*. The same concentration of nanoparticles in solution would have yielded a PARACEST contrast-to-noise ratio of only 7.99, but the clots reached a contrast-to-noise ratio of 17.7. Based on the ^{19}F quantitation, the detection limit (contrast-to-noise ratio = 5) for bound PARACEST nanoparticles was 2.30 n*M,* which is 44% lower than the detection limit for nanoparticles in solution. These results suggest that

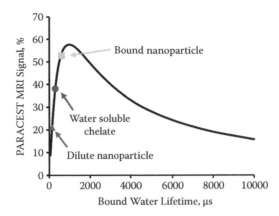

FIGURE 3.16 Modeling of the Bloch equations provides an estimation of the PARACEST MR signal for agents with different bound water lifetimes. The bound water lifetime for the PFC nanoparticles in solution (triangle) is lower than for the water-soluble chelate (circle), decreasing the observed PARACEST contrast. When the particles bind to a target surface, such as the fibrin clots, the bound water lifetime increases (square). The increased bound water lifetime is closer to the optimal exchange rate, causing the PARACEST contrast to increase. (Reprinted with permission from Cai, K. et al. 2011. *NMR Biomedicine* 25:279–285.)

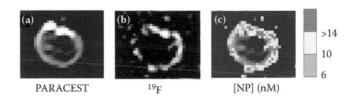

FIGURE 3.17 **(See color insert.)** Combined PARACEST and ^{19}F molecular imaging of fibrin-targeted PARACEST PFC nanoparticles. PARACEST (a) and ^{19}F (b) images both reveal nanoparticles bound to the clot surface. (c) The ^{19}F signal is converted to nanoparticle concentration (nM) and color coded in an overlay on the PARACEST subtraction image, demonstrating colocalization of these two definitive signals. (Reprinted with permission from Cai, K. et al. 2011. *NMR Biomedicine* 25:279–285.)

binding the nanoparticles onto a biological target actually improves the PARACEST mechanism, perhaps due to an increase in the bound water lifetime. Mathematical simulation of the Bloch equations indicates that the bound water lifetime of the bound nanoparticles was around 600 μs (Figure 3.16), which is much closer to the optimum value than nanoparticles in solution (108 μs). This study demonstrates that PARACEST PFC nanoparticles provide a unique opportunity for dual PARACEST and ^{19}F MR molecular imaging, to corroborate specific binding of the particles to the target, and for quantitation of particles within a voxel. These properties allow optimization of the chelate chemistry and MR imaging pulse sequence parameters to achieve the maximum image contrast.

3.7 CONCLUSION

In conclusion, there has been a recent upwelling of research and development of nanoparticle systems for a variety of applications, including molecular imaging, drug delivery, cell labeling, and other biological uses. The physical and chemical properties of nanoparticles can vary widely as their basic structure may consist of dendrimers, viruses, liposomes, micelles, PFC particles, etc. The ability to modify the surface and/or the core components of the particles allows them to be engineered specifically to bind to cellular biomarkers, target distinctive cell populations, produce a reliable MR imaging signal, transport potent pharmaceuticals directly to the diseased tissue, and/or release therapeutics in response to an external trigger. Combining the roles of diagnosis and therapy into a single agent, these theranostic nanoparticles could revolutionize the detection and treatment of many of the most critical clinical diseases, including cancer, stroke, cardiovascular disease, obesity, diabetes, and others. While the complexity and manufacturing hurdles for these multipurpose agents will certainly delay their adoption for widespread clinical use, the advantages and proposed benefits of nanomedicine will continue to spur intense research efforts to demonstrate new and exciting biological capabilities of nanoparticles.

REFERENCES

1. Cromer Berman, S. M., P. Walczak, and J. W. Bulte. 2011. Tracking stem cells using magnetic nanoparticles. *Wiley Interdisciplinary Reviews in Nanomedicine and Nanobiotechnology* 3 (4): 343–355.
2. Kanwar, R. K. et al. 2012. Emerging engineered magnetic nanoparticulate probes for targeted MRI of atherosclerotic plaque macrophages. *Nanomedicine* (London) 7 (5): 735–749.
3. Lee, N. and T. Hyeon. 2012. Designed synthesis of uniformly sized iron oxide nanoparticles for efficient magnetic resonance imaging contrast agents. *Chemical Society Reviews* 41 (7): 2575–2589.
4. Tassa, C., S. Y. Shaw, and R. Weissleder. 2011. Dextran-coated iron oxide nanoparticles: A versatile platform for targeted molecular imaging, molecular diagnostics, and therapy. *Accounts of Chemistry Research* 44 (10): 842–852.
5. Tu, C. and A. Y. Louie. 2012. Nanoformulations for molecular MRI. *Wiley Interdisciplinary Reviews in Nanomedicine and Nanobiotechnology* 4 (4): 448–457.
6. Pan, D. et al. 2010. Nanomedicine strategies for molecular targets with MRI and optical imaging. *Future Medicinal Chemistry* 2 (3): 471–490.
7. Winter, P. M. et al. 2010. Quantitative cardiovascular magnetic resonance for molecular imaging. *Journal of Cardiovascular Magnetic Resonance* 12: 62.
8. Chen, J., G. M. Lanza, and S. A. Wickline. 2010. Quantitative magnetic resonance fluorine imaging: Today and tomorrow. *Wiley Interdisciplinary Reviews in Nanomedicine and Nanobiotechnology* 2 (4): 431–440.
9. Horton, M. A. 1997. The alpha v beta 3 integrin "vitronectin receptor." *International Journal of Biochemical Cell Biology* 29 (5): 721–725.
10. Lanza, G. M. et al. 2000. In vivo molecular imaging of stretch-induced tissue factor in carotid arteries with ligand-targeted nanoparticles. *Journal of American Society of Echocardiography* 13 (6): 608–614.

11. Winter, P. M. et al. 2005. Molecular imaging of human thrombus with computed tomography. *Academic Radiology* 12 (Suppl 1): S9–13.
12. Lijowski, M. et al. 2009. High sensitivity: High-resolution SPECT-CT/MR molecular imaging of angiogenesis in the Vx2 model. *Investigations in Radiology* 44 (1): 15–22.
13. Schmieder, A. H. et al. 2008. Three-dimensional MR mapping of angiogenesis with alpha5beta1(alpha nu beta3)-targeted theranostic nanoparticles in the MDA-MB-435 xenograft mouse model. *FASEB Journal* 22 (12): 4179–4189.
14. Winter, P. M. et al. 2008. Antiangiogenic synergism of integrin-targeted fumagillin nanoparticles and atorvastatin in atherosclerosis. *JACC Cardiovascular Imaging* 1 (5): 624–634.
15. Zhou, H. F., et al. 2009. Alphavbeta3-targeted nanotherapy suppresses inflammatory arthritis in mice. *FASEB Journal* 23 (9): 2978–2985.
16. Stanisz, G. J. and R. M. Henkelman. 2000. Gd-DTPA relaxivity depends on macromolecular content. *Magnetic Resonance Medicine* 44 (5): 665–667.
17. Morawski, A. M. et al. 2004. Targeted nanoparticles for quantitative imaging of sparse molecular epitopes with MRI. *Magnetic Resonance Medicine* 51 (3): 480–486.
18. Winter, P. M. et al. 2003. Molecular imaging of angiogenesis in nascent Vx-2 rabbit tumors using a novel alpha(nu)beta3-targeted nanoparticle and 1.5 tesla magnetic resonance imaging. *Cancer Research* 63 (18): 5838–5843.
19. Winter, P. M. et al. 2008. Minute dosages of alpha(nu)beta3-targeted fumagillin nanoparticles impair Vx-2 tumor angiogenesis and development in rabbits. *FASEB Journal* 22 (8): 2758–2767.
20. Winter, P. M. et al. 2003. Molecular imaging of angiogenesis in early stage atherosclerosis with alpha(v)beta3-integrin-targeted nanoparticles. *Circulation* 108 (18): 2270–2274.
21. Cai, K. et al. 2010. MR molecular imaging of aortic angiogenesis. *JACC Cardiovascular Imaging* 3 (8): 824–832.
22. Winter, P. M. et al. 2010. Molecular imaging of angiogenic therapy in peripheral vascular disease with alphanubeta3-integrin-targeted nanoparticles. *Magnetic Resonance Medicine* 64 (2): 369–376.
23. Winter, P. M. et al. 2006. Endothelial alpha(v)beta3 integrin-targeted fumagillin nanoparticles inhibit angiogenesis in atherosclerosis. *Arteriosclerosis, Thrombosis and Vascular Biology* 26 (9): 2103–2109.
24. Southworth, R. et al. 2009. Renal vascular inflammation induced by Western diet in ApoE-null mice quantified by (19)F NMR of VCAM-1 targeted nanobeacons. *Nanomedicine* 5 (3): 359–367.
25. Boles, K. S. et al. 2010. MR angiogenesis imaging with Robo4- vs. alphaVbeta3-targeted nanoparticles in a B16/F10 mouse melanoma model. *FASEB Journal* 24 (11): 4262–4270.
26. Waters, E. A. et al. 2008. Detection of targeted perfluorocarbon nanoparticle binding using 19F diffusion weighted MR spectroscopy. *Magnetic Resonance Medicine* 60 (5): 1232–1236.
27. Morawski, A. M. et al. 2004. Quantitative "magnetic resonance immunohistochemistry" with ligand-targeted (19)F nanoparticles. *Magnetic Resonance Medicine* 52 (6): 1255–1262.
28. Caruthers, S. D. et al. 2006. In vitro demonstration using 19F magnetic resonance to augment molecular imaging with paramagnetic perfluorocarbon nanoparticles at 1.5 tesla. *Investigative Radiology* 41 (3): 305–312.
29. Waters, E. A. et al. 2008. Detection and quantification of angiogenesis in experimental valve disease with integrin-targeted nanoparticles and 19-fluorine MRI/MRS. *Journal of Cardiovascular Magnetic Resonance* 10 (1): 43.

30. Hu, L. et al. 2013. Rapid quantification of oxygen tension in blood flow with a fluorine nanoparticle reporter and a novel blood flow-enhanced-saturation-recovery sequence. *Magnetic Resonance Medicine* 70:176–183.
31. Bangham, A. D. and R. W. Horne. 1964. Negative staining of phospholipids and their structural modification by surface-active agents as observed in the electron microscope. *Journal of Molecular Biology* 8:660–668.
32. Beija, M. et al. 2012. Colloidal systems for drug delivery: From design to therapy. *Trends in Biotechnology* 30 (9): 485–496.
33. Strijkers, G. J. et al. 2010. Paramagnetic and fluorescent liposomes for target-specific imaging and therapy of tumor angiogenesis. *Angiogenesis* 13 (2): 161–173.
34. Sipkins, D. A. et al. 1998. Detection of tumor angiogenesis in vivo by alphaVbeta3-targeted magnetic resonance imaging. *Nature Medicine* 4 (5): 623–626.
35. Zhang, Y. et al. 2011. Multimodality molecular imaging of CD105 (Endoglin) expression. *International Journal of Clinical and Experimental Medicine* 4 (1): 32–42.
36. Nassiri, F. et al. 2011. Endoglin (CD105): A review of its role in angiogenesis and tumor diagnosis, progression and therapy. *Anticancer Research* 31 (6): 2283–2290.
37. Maiseyeu, A. et al. 2009. Gadolinium-containing phosphatidylserine liposomes for molecular imaging of atherosclerosis. *Journal of Lipid Research* 50 (11): 2157–2163.
38. Liu, F. et al. 2011. Superparamagnetic nanosystems based on iron oxide nanoparticles for biomedical imaging. *Nanomedicine* (Lond) 6 (3): 519–528.
39. Nahrendorf, M. et al. 2006. Noninvasive vascular cell adhesion molecule-1 imaging identifies inflammatory activation of cells in atherosclerosis. *Circulation* 114 (14): 1504–1511.
40. Rosenblum, L. T. et al. 2010. In vivo molecular imaging using nanomaterials: General in vivo characteristics of nano-sized reagents and applications for cancer diagnosis. *Molecular Membrane Biology* 27 (7): 274–285.
41. Sillerud, L. O. et al. 2013. SPION-enhanced magnetic resonance imaging of Alzheimer's disease plaques in AbetaPP/PS-1 transgenic mouse brain. *Journal of Alzheimers Disease* 34 (2): 349–365.
42. Andreas, K. et al. 2012. Highly efficient magnetic stem cell labeling with citrate-coated superparamagnetic iron oxide nanoparticles for MRI tracking. *Biomaterials* 33 (18): 4515–4525.
43. Bulte, J. W. 2009. In vivo MRI cell tracking: Clinical studies. *American Journal of Roentgenology* 193 (2): 314–325.
44. Link, T. W. et al. 2012. Use of magnetocapsules for in vivo visualization and enhanced survival of xenogeneic HepG2 cell transplants. *Cell Media* 4 (2): 77–84.
45. Odintsov, B. et al. 2011. 14.1 T whole body MRI for detection of mesoangioblast stem cells in a murine model of Duchenne muscular dystrophy. *Magnetic Resonance Medicine* 66 (6): 1704–1714.
46. Qi, X. and G. A. Grabowski 2001. Differential membrane interactions of saposins A and C: Implications for the functional specificity. *Journal of Biological Chemistry* 276 (29): 27010–27017.
47. Qi, X. et al. 1996. Functional organization of saposin C. Definition of the neurotrophic and acid beta-glucosidase activation regions. *Journal of Biological Chemistry* 271 (12): 6874–6880.
48. Ran, S. and P. E. Thorpe. 2002. Phosphatidylserine is a marker of tumor vasculature and a potential target for cancer imaging and therapy. *International Journal of Radiation Oncology Biology Physics* 54 (5): 1479–1484.
49. Utsugi, T. et al. 1991. Elevated expression of phosphatidylserine in the outer membrane leaflet of human tumor cells and recognition by activated human blood monocytes. *Cancer Research* 51 (11): 3062–3066.
50. Kaimal, V. et al. 2011. Saposin C coupled lipid nanovesicles enable cancer-selective optical and magnetic resonance imaging. *Molecular Imaging Biology* 13 (5): 886–897.

51. Qi, X. et al. 2009. Cancer-selective targeting and cytotoxicity by liposomal-coupled lysosomal saposin C protein. *Clinical Cancer Research* 15 (18): 5840–5851.
52. Bogdanov, A. A., Jr., et al. 1994. Trapping of dextran-coated colloids in liposomes by transient binding to aminophospholipid: Preparation of ferrosomes. *Biochimica et Biophysica Acta* 1193 (1): 212–218.
53. Mantripragada, S. 2002. A lipid based depot (DepoFoam technology) for sustained release drug delivery. *Progress in Lipid Research* 41 (5): 392–406.
54. Verma, S. et al. 2008. Metastatic breast cancer: The role of pegylated liposomal doxorubicin after conventional anthracyclines. *Cancer Treatment Reviews* 34 (5): 391–406.
55. Decker, C. et al. 2013. Pharmacokinetics of temoporfin-loaded liposome formulations: Correlation of liposome and temoporfin blood concentration. *Journal of Control Release* 166 (3): 277–285.
56. Yang, X. et al. 2009. A novel liposomal formulation of flavopiridol. *International Journal of Pharmaceutics* 365 (1–2): 170–174.
57. de Smet, M. et al. 2011. Magnetic resonance imaging of high intensity focused ultrasound mediated drug delivery from temperature-sensitive liposomes: An in vivo proof-of-concept study. *Journal of Control Release* 150 (1): 102–110.
58. Tagami, T. et al. 2011. MRI monitoring of intratumoral drug delivery and prediction of the therapeutic effect with a multifunctional thermosensitive liposome. *Biomaterials* 32 (27): 6570–6578.
59. Frias, J. C. et al. 2004. Recombinant HDL-like nanoparticles: A specific contrast agent for MRI of atherosclerotic plaques. *Journal of American Chemical Society* 126 (50): 16316–16317.
60. Corbin, I. R. et al. 2006. Low-density lipoprotein nanoparticles as magnetic resonance imaging contrast agents. *Neoplasia* 8 (6): 488–498.
61. Mulder, W. J. et al. 2009. Molecular imaging of tumor angiogenesis using alphavbeta3-integrin targeted multimodal quantum dots. *Angiogenesis* 12 (1): 17–24.
62. Allen, M. et al. 2005. Paramagnetic viral nanoparticles as potential high-relaxivity magnetic resonance contrast agents. *Magnetic Resonance Medicine* 54 (4): 807–812.
63. Singh, P. et al. 2007. Bio-distribution, toxicity and pathology of cowpea mosaic virus nanoparticles in vivo. *Journal of Control Release* 120 (1–2): 41–50.
64. Anderson, E. A. et al. 2006. Viral nanoparticles donning a paramagnetic coat: Conjugation of MRI contrast agents to the MS2 capsid. *Nano Letters* 6 (6): 1160–1164.
65. Vasalatiy, O. et al. 2008. Labeling of adenovirus particles with PARACEST agents. *Bioconjugate Chemistry* 19 (3): 598–606.
66. Senger, D. R. et al. 1983. Tumor cells secrete a vascular permeability factor that promotes accumulation of ascites fluid. *Science* 219 (4587): 983–985.
67. He, T. et al. 2011. Molecular MRI assessment of vascular endothelial growth factor receptor-2 in rat C6 gliomas. *Journal of Cellular and Molecular Medicine* 15 (4): 837–849.
68. Liu, Y. et al. 2011. Gadolinium-loaded polymeric nanoparticles modified with anti-VEGF as multifunctional MRI contrast agents for the diagnosis of liver cancer. *Biomaterials* 32 (22): 5167–5176.
69. Qiao, J. et al. 2011. HER2 targeted molecular MR imaging using a de novo designed protein contrast agent. *PLoS One* 6 (3): e18103.
70. Zhu, W. et al. 2008. PAMAM dendrimer-based contrast agents for MR imaging of Her-2/neu receptors by a three-step pretargeting approach. *Magnetic Resonance Medicine* 59 (4): 679–685.
71. Buxton, D. B. et al. 2011. Report of the National Heart, Lung, and Blood Institute working group on the translation of cardiovascular molecular imaging. *Circulation* 123 (19): 2157–2163.
72. Gaemperli, O. and P. A. Kaufmann. 2011. PET and PET/CT in cardiovascular disease. *Annals of New York Academy of Sciences* 1228:109–136.

73. Gulyas, B. and C. Halldin. 2012. New PET radiopharmaceuticals beyond FDG for brain tumor imaging. *Quarterly Journal of Nuclear Medicine and Molecular Imaging* 56 (2): 173–190.
74. Jadvar, H. 2009. Molecular imaging of prostate cancer with 18F-fluorodeoxyglucose PET. *National Review of Urology* 6 (6): 317–323.
75. Langsteger, W. et al. 2011. Fluorocholine (18F) and sodium fluoride (18F) PET/CT in the detection of prostate cancer: Prospective comparison of diagnostic performance determined by masked reading. *Quarterly Journal of Nuclear Medicine and Molecular Imaging* 55 (4): 448–457.
76. Stacy, M. R., M. W. Maxfield, and A. J. Sinusas. 2012. Targeted molecular imaging of angiogenesis in PET and SPECT: A review. *Yale Journal of Biological Medicine* 85 (1): 75–86.
77. Wagner, S. and K. Kopka. 2013. Non-peptidyl (18)F-labelled PET tracers as radio-indicators for the noninvasive detection of cancer. *Recent Results in Cancer Research* 187:107–132.
78. Parmelee, D. J. et al. 1997. Preclinical evaluation of the pharmacokinetics, biodistribution, and elimination of MS-325, a blood pool agent for magnetic resonance imaging. *Investigative Radiology* 32 (12): 741–747.
79. Hartmann, M. et al. 2006. Initial imaging recommendations for Vasovist angiography. *European Radiology* 16 (Suppl 2): B15–23.
80. Lewis, M., S. Yanny, and P. N. Malcolm. 2012. Advantages of blood pool contrast agents in MR angiography: A pictorial review. *Journal of Medical Imaging Radiation Oncology* 56 (2): 187–191.
81. Overoye-Chan, K. et al. 2008. EP-2104R: A fibrin-specific gadolinium-based MRI contrast agent for detection of thrombus. *Journal of American Chemical Society* 130 (18): 6025–6039.
82. Spuentrup, E. et al. 2008. MR imaging of thrombi using EP-2104R, a fibrin-specific contrast agent: Initial results in patients. *European Radiology* 18 (9): 1995–2005.
83. Vymazal, J. et al. 2009. Thrombus imaging with fibrin-specific gadolinium-based MR contrast agent EP-2104R: Results of a phase II clinical study of feasibility. *Investigative Radiology* 44 (11): 697–704.
84. Spuentrup, E. et al. 2005. Molecular magnetic resonance imaging of coronary thrombosis and pulmonary emboli with a novel fibrin-targeted contrast agent. *Circulation* 111 (11): 1377–1382.
85. Stracke, C. P. et al. 2007. Molecular MRI of cerebral venous sinus thrombosis using a new fibrin-specific MR contrast agent. *Stroke* 38 (5): 1476–1481.
86. Zhang, S. et al. 2001. A novel europium(III)-based MRI contrast agent. *Journal of American Chemical Society* 123 (7): 1517–1518.
87. Ward, K. M., A. H. Aletras, and R. S. Balaban. 2000. A new class of contrast agents for MRI based on proton chemical exchange dependent saturation transfer (CEST). *Journal of Magnetic Resonance* 143 (1): 79–87.
88. Aime, S. et al. 2002. Paramagnetic lanthanide(III) complexes as pH-sensitive chemical exchange saturation transfer (CEST) contrast agents for MRI applications. *Magnetic Resonance Medicine* 47 (4): 639–648.
89. Terreno, E. et al. 2004. Ln(III)-DOTAMGly complexes: A versatile series to assess the determinants of the efficacy of paramagnetic chemical exchange saturation transfer agents for magnetic resonance imaging applications. *Investigative Radiology* 39 (4): 235–243.
90. Zhang, S., C. R. Malloy, and A. D. Sherry. 2005. MRI thermometry based on PARACEST agents. *Journal of American Chemical Society* 127 (50): 17572–17573.
91. Aime, S. et al. 2002. A paramagnetic MRI-CEST agent responsive to lactate concentration. *Journal of American Chemical Society* 124 (32): 9364–9365.

92. Trokowski, R., S. Zhang, and A. D. Sherry. 2004. Cyclen-based phenylboronate ligands and their Eu3+ complexes for sensing glucose by MRI. *Bioconjugate Chemistry* 15 (6): 1431–1440.
93. Trokowski, R. et al. 2005. Selective sensing of zinc ions with a PARACEST contrast agent. *Angewandte Chemie* International Edition England 44 (42): 6920–6923.
94. Aime, S., D. Delli Castelli, and E. Terreno. 2005. Highly sensitive MRI chemical exchange saturation transfer agents using liposomes. *Angewandte Chemie* International Edition England 44 (34): 5513–5515.
95. Aime, S. et al. 2007. Gd-loaded liposomes as T1, susceptibility, and CEST agents, all in one. *Journal of American Chemical Society* 129 (9): 2430–2431.
96. Terreno, E. et al. 2007. From spherical to osmotically shrunken paramagnetic liposomes: An improved generation of LIPOCEST MRI agents with highly shifted water protons. *Angewandte Chemie* International Edition England 46 (6): 966–968.
97. Delli Castelli, D. et al. 2008. Lanthanide-loaded paramagnetic liposomes as switchable magnetically oriented nanovesicles. *Inorganic Chemistry* 47 (8): 2928–2930.
98. Terreno, E. et al. 2008. Highly shifted LIPOCEST agents based on the encapsulation of neutral polynuclear paramagnetic shift reagents. *Chemistry Communications* (Cambridge) (5): 600–602.
99. Flament, J. et al. 2013. In vivo CEST MR imaging of U87 mice brain tumor angiogenesis using targeted LipoCEST contrast agent at 7 T. *Magnetic Resonance Medicine* 69 (1): 179–187.
100. Ali, M. M., B. Yoo, and M. D. Pagel 2009. Tracking the relative in vivo pharmacokinetics of nanoparticles with PARACEST MRI. *Molecular Pharmacology* 6 (5): 1409–1416.
101. Winter, P. M. et al. 2006. Targeted PARACEST nanoparticle contrast agent for the detection of fibrin. *Magnetic Resonance Medicine* 56 (6): 1384–1388.
102. Cai, K. et al. 2011. Quantification of water exchange kinetics for targeted PARACEST perfluorocarbon nanoparticles. *NMR Biomedicine* 25:279–285.
103. Liu, G. et al. 2011. Imaging in vivo extracellular pH with a single paramagnetic chemical exchange saturation transfer magnetic resonance imaging contrast agent. *Molecular Imaging* 11:47–57.
104. Sheth, V. R. et al. 2012. Measuring in vivo tumor pHe with CEST-FISP MRI. *Magnetic Resonance Medicine* 67:760–768.
105. Li, Y. et al. 2011. A self-calibrating PARACEST MRI contrast agent that detects esterase enzyme activity. *Contrast Media Molecular Imaging* 6 (4): 219–228.
106. Liu, G., Y. Li, and M. D. Pagel. 2007. Design and characterization of a new irreversible responsive PARACEST MRI contrast agent that detects nitric oxide. *Magnetic Resonance Medicine* 58 (6): 1249–1256.
107. Yoo, B. and M. D. Pagel. 2006. A PARACEST MRI contrast agent to detect enzyme activity. *Journal of American Chemical Society* 128 (43): 14032–14033.
108. Winter, P. M. et al. 2003. Improved molecular imaging contrast agent for detection of human thrombus. *Magnetic Resonance Medicine* 50 (2): 411–416.
109. Flacke, S. et al. 2001. Novel MRI contrast agent for molecular imaging of fibrin: Implications for detecting vulnerable plaques. *Circulation* 104 (11): 1280–1285.
110. Neubauer, A. M. et al. 2008. Gadolinium-modulated 19F signals from perfluorocarbon nanoparticles as a new strategy for molecular imaging. *Magnetic Resonance Medicine* 60 (5): 1066–1072.

- Humans are 10,000,000 times smaller than the Earth.
- A 100 nm sized particle, is 10,000,000 times smaller than a human.

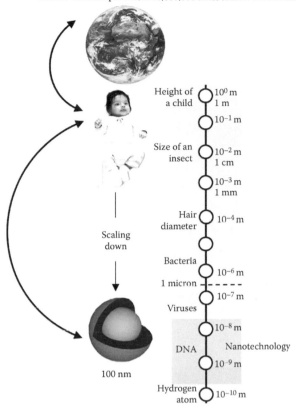

Height of a child	10^0 m / 1 m
	10^{-1} m
Size of an insect	10^{-2} m / 1 cm
	10^{-3} m / 1 mm
Hair diameter	10^{-4} m
Bacteria	10^{-6} m
1 micron	10^{-7} m
Viruses	
	10^{-8} m
DNA — Nanotechnology	
	10^{-9} m
Hydrogen atom	10^{-10} m

Scaling down

100 nm

FIGURE 1.1 Size scales in nanotechnology.

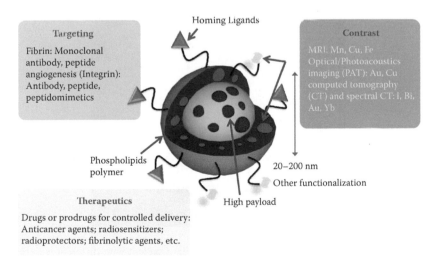

Homing Ligands

Targeting

Fibrin: Monoclonal antibody, peptide angiogenesis (Integrin): Antibody, peptide, peptidomimetics

Contrast

MRI: Mn, Cu, Fe Optical/Photoacoustics imaging (PAT): Au, Cu computed tomography (CT) and spectral CT: I, Bi, Au, Yb

Phospholipids polymer

20–200 nm

Other functionalization

Therapeutics

Drugs or prodrugs for controlled delivery: Anticancer agents; radiosensitizers; radioprotectors; fibrinolytic agents, etc.

High payload

FIGURE 1.2 A cartoon of multifunctional nanoparticles representing an extremely versatile platform for molecular imaging and drug delivery application.

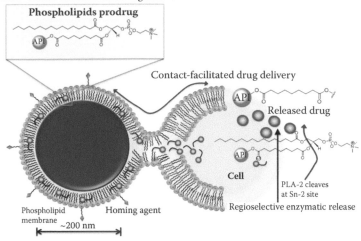

API= Active Pharmaceutics Ingredient

Phospholipids prodrug

Contact-facilitated drug delivery

API

Released drug

Cell

PLA-2 cleaves
at Sn-2 site

Phospholipid
membrane

Homing agent

Regioselective enzymatic release

~200 nm

FIGURE 1.3 A cartoon depicting the mechanistic concept of a therapeutic delivery via "contact facilitated drug delivery" mechanism.

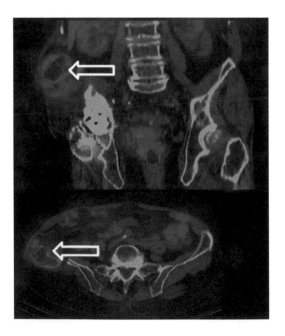

FIGURE 2.2 SPECT/CT imaging with ^{131}I-iodide in a patient with differentiated follicular thyroid carcinoma after pelvic surgery shows a benign tracer accumulation in the colon (arrow). (Adapted with permission from Mariani, G. et al. 2010. *European Journal of Nuclear Medicine and Molecular Imaging* 10:1959–1985.)

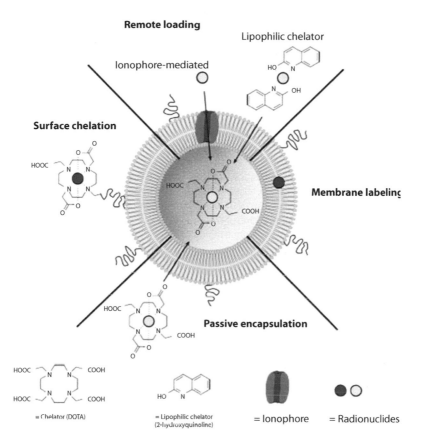

FIGURE 2.6 Schematic diagram of the remote loading, membrane labeling, passive encapsulation, and surface chelation methods for preparing radioactive liposomes. Radionuclides can be associated with the lipid membrane by hydrophobic interaction, through membrane conjugation, or surface chelation using chelator–lipid conjugates in preformed liposomes (*blue radionuclides*). Radionuclides can alternatively be encapsulated inside liposomes during lipid hydration or can be transported through the lipid membrane of preformed liposomes by ionophores or lipophilic chelators (*yellow radionuclides*). In the latter case, the radionuclides are trapped inside the aqueous lumen by a hydrophilic chelator with high affinity for the radionuclide. (Reproduced with permission from Petersen, A. L. et al. 2012. *Advanced Drug Delivery Reviews* 13:1417–1435.)

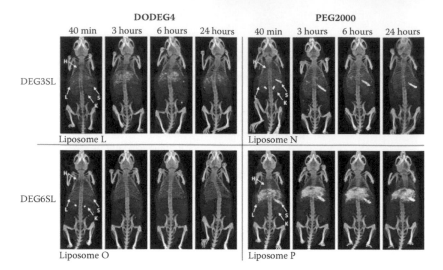

FIGURE 2.7 SPECT/CT images showing that both of the control PEG2000 liposomes (N and P) appear to show a characteristically high uptake in the spleen compared to both DODEG4 liposomes (left) that possess a lipid multifunctional, multimodal shielded liposomal formulation. Images were acquired over 24 hr (H—heart, L—liver, S—spleen, and K—kidney). (Reproduced with permission from Mitchell, N. et al. 2013. *Biomaterials* 4:1179–1192.)

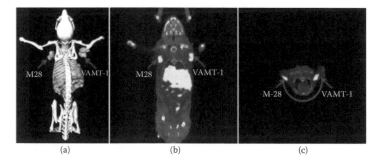

FIGURE 2.8 SPECT/CT fused image of ^{111}In-IL-M1$_{(50)}$ taken 24 h after injection; (a) three-dimensional reconstruction, (b) coronal view, (c) transverse view. The uptake of ^{111}In-IL-M1$_{(50)}$ in both epithelioid (M28) and sarcomatoid (VAMT-1) mesothelioma tumors at 24 hr was shown. (Reproduced with permission from Iyer, A. K. et al. 2011. *Biomaterials* 10:2605–2613.)

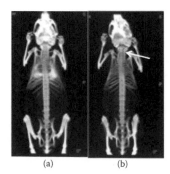

FIGURE 2.9 Whole-body SPECT/CT imaging in mice at 7 d postinjection showing (a) persistent lung accumulation of GlcNAc^D–Na^{125}I@SWNTs and (b) thyroid accumulation of free Na^{125}I (unencapsulated), indicating effective and complete encapsulation of radionuclide within the nanocapsule and long-term stability of the construct. (Adapted with permission from Hong, S. Y. et al. 2010. *Nature Materials* 6:485–490.)

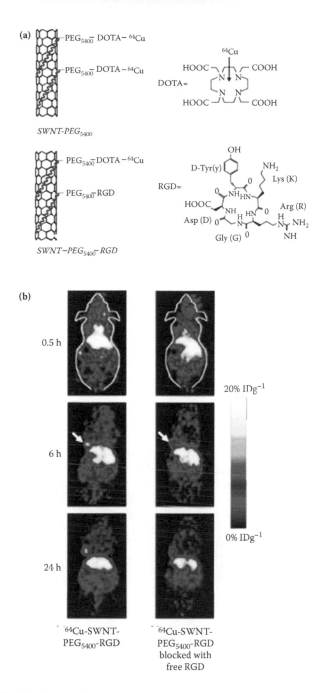

FIGURE 2.10 (a) Schematic drawings of noncovalently functionalized SWNT-PEG$_{5400}$-RGD conjugates with ^{64}Cu-DOTA. The hydrophobic carbon chains (blue segments) of the phospholipids strongly bind to the sidewalls of the SWNTs, and the PEG chains render water solubility to the SWNTs. (b) MicroPET images of U87MG tumor-bearing mice showing high tumor uptake of ^{64}Cu-SWNT-PEG$_{5400}$-RGD, which was significantly reduced by co-injection of free RGD peptide. (Adapted with permission from Liu, Z. et al. 2007. *Nature Nanotechnology* 1:47–52.)

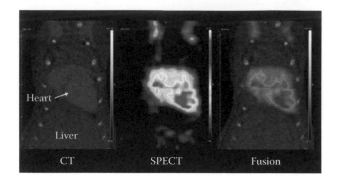

FIGURE 2.12 Multimodal imaging of the G4-[[[[Ac]-TIBA]-DTPA]-mPEG$_{12}$] dendrimer construct in normal mice by microCT and microSPECT. The fusion image shows significant colocalization of the nuclear and x-ray contrast components. (Reproduced with permission from Criscione, J. M. et al. 2011. *Bioconjugate Chemistry* 9:1784–1792.)

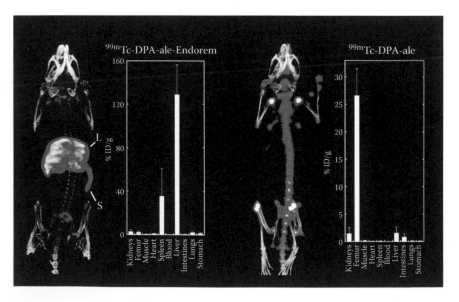

FIGURE 2.15 Whole-body SPECT/CT images and biodistribution studies of (a) a ^{99m}Tc-labeled SPIO-bisphosphonate construct and (b) a conventional ^{99m}Tc-bisphosphonate agent. (Reproduced with permission from Torres Martin de Rosales, R. et al. 2011. *Bioconjugate Chemistry* 3:455–465.)

(a)

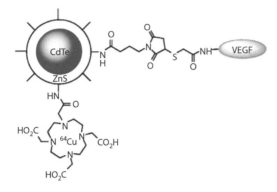

(b)

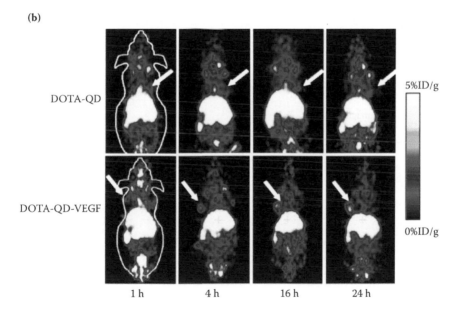

FIGURE 2.16 (a) Structure of DOTA-QD-VEGF conjugate. (b) Whole-body coronal PET images of U87MG tumor-bearing mice at 1, 4, 16, and 24 hr after injection of ^{64}Cu-DOTA-QD and ^{64}Cu-DOTA-QD-VEGF. Arrows indicate the tumor. (Adapted with permission from Chen, K. et al. 2008. *European Journal of Nuclear Medicine and Molecular Imaging* 12:2235–2244.)

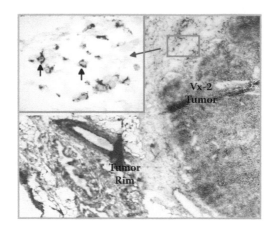

FIGURE 3.2 Histology of rabbit tumor showing angiogenesis (black arrows) near a large blood vessel adjacent to the tumor rim. Tumor sections were stained with H&E (low magnification image) to show morphology or LM-609 (inset, high magnification image) to show $\alpha_v\beta_3$-integrin expression. The anatomical location of angiogenic vasculature determined via histology corresponds to the areas of MR signal enhancement following injection of $\alpha_v\beta_3$-integrin-targeted nanoparticles. (Reprinted with permission from Winter, P. M. et al. 2003. *Cancer Research* 63 (18): 5838–5843.)

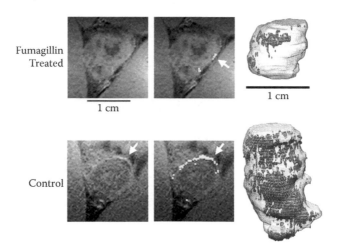

FIGURE 3.3 Reduced MR contrast enhancement in T1-weighted images of a rabbit treated with $\alpha_v\beta_3$-integrin-targeted fumagillin nanoparticles (top) compared to an animal receiving $\alpha_v\beta_3$-integrin-targeted nanoparticles without drug (bottom). Enhancing pixels, color coded in yellow (white arrows), demonstrate large areas of angiogenesis in the control tumor and markedly lower levels of angiogenesis with fumagillin treatment. Panels on the right-hand side demonstrate 3D neovascular maps of the tumors with and without fumagillin therapy. The angiogenic pixels are color coded in blue and are much more prevalent in the periphery of the control tumor than the treated tumor. (Reprinted with permission from Winter, P. M. et al. 2008. *FASEB Journal* 22 (8): 2758–2767.)

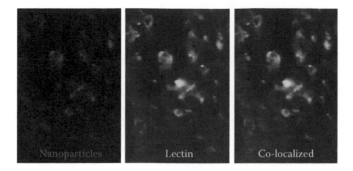

FIGURE 3.4 Fluorescence microscopy (20 times magnification) of the tumor periphery showing $\alpha_v\beta_3$-integrin-targeted nanoparticles containing rhodamine (left) and FITC-lectin (middle). Overlaying the fluorescent signal from these two agents demonstrates that $\alpha_v\beta_3$-integrin-targeted nanoparticles are constrained to the vasculature and taken up by angiogenic capillaries. (Reprinted with permission from Winter, P. M. et al. 2008. *FASEB Journal* 22 (8): 2758–2767.)

Molecular imaging of angiogenesis

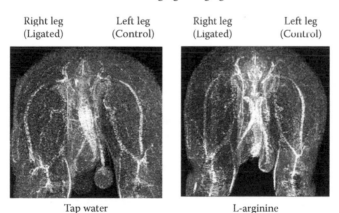

FIGURE 3.9 MR molecular imaging enhancement (color coded in red) shows diffuse angiogenesis throughout the ischemic leg (right) with only slight enhancement in the control leg (left). The animal receiving tap water (left panel) shows more enhancement in the ligated leg compared to the control leg, but L-arginine treatment (right panel) results in a more dense distribution of angiogenesis in the ligated limb, while the control limb appears similar to the untreated animal. (Reprinted with permission from Winter, P. M. et al. 2010. *Magnetic Resonance Medicine* 64 (2): 369–376.)

H&E staining CD31 staining
Tap water Tap water L-arginine

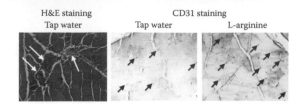

FIGURE 3.10 Histology of muscle from the ischemic limb of animals treated with tap water or L-arginine. Left: Hematoxylin and eosin (H&E) staining shows intramuscular hemorrhage (white arrows) in tap water animals, which was not observed with L-arginine treatment. Middle and right: Microvascular staining (black arrows) revealed more capillaries in L-arginine-treated animals compared to animals receiving tap water, supporting the MR molecular imaging and x-ray angiography findings that L-arginine augments angiogenic response to limb ischemia. (Reprinted with permission from Winter, P. M. et al. 2010. *Magnetic Resonance Medicine* 64 (2): 369–376.)

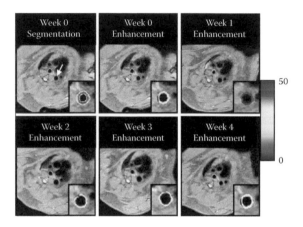

FIGURE 3.13 Serial MR molecular imaging of angiogenesis in the aortic wall before and after targeted fumagillin treatment. In the baseline image (week 0 panels), cross-sectional imaging of the thoracic aorta (arrow) provides robust segmentation of the aortic wall (yellow outline) and reveals patchy areas of angiogenesis (color-coded overlay). Following treatment (week 1 panel), the signal enhancement is markedly lower due to the antiangiogenic effect of fumagillin. The level of signal enhancement gradually increases (week 2 and week 3 panels) and eventually returns to the pretreatment value (week 4 panel). (Reprinted with permission from Cai, K. et al. 2010. *JACC Cardiovascular Imaging* 3 (8): 824–832.)

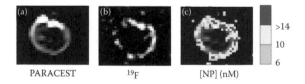

PARACEST ^{19}F [NP] (nM)

FIGURE 3.17 Combined PARACEST and ^{19}F molecular imaging of fibrin-targeted PARACEST PFC nanoparticles. PARACEST (a) and ^{19}F (b) images both reveal nanoparticles bound to the clot surface. (c) The ^{19}F signal is converted to nanoparticle concentration (nM) and color coded in an overlay on the PARACEST subtraction image, demonstrating colocalization of these two definitive signals. (Reprinted with permission from Cai, K. et al. 2011. *NMR Biomedicine* 25:279–285.)

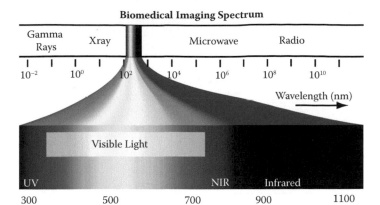

FIGURE 4.1 Comparative energy spectra and location of non-ionizing light in the electromagnetic spectrum.

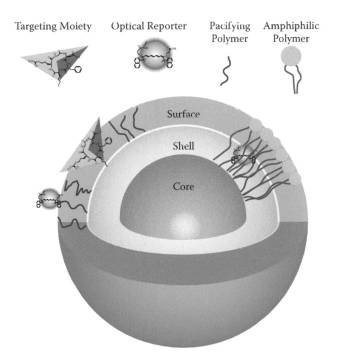

FIGURE 4.2 Simplified cartoon of nanoparticles contrast agent platforms for optical imaging. Nanoparticles are described according to the number of material layers and chemical and biochemical modifications to pacify their surfaces to escape innate immune responses and improve distribution characteristics.

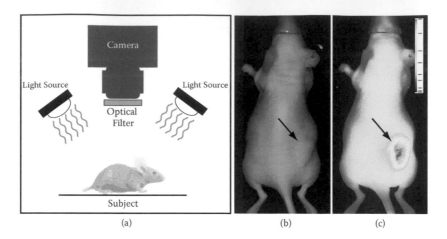

FIGURE 4.4 (a) Example planar reflectance imaging system setup for detection of fluorescence in mouse cancer model. (b) Brightfield and (c) fluorescence images of mouse after intravenous administration of tumor-targeted molecular probe in mouse with subcutaneous tumor (arrow).

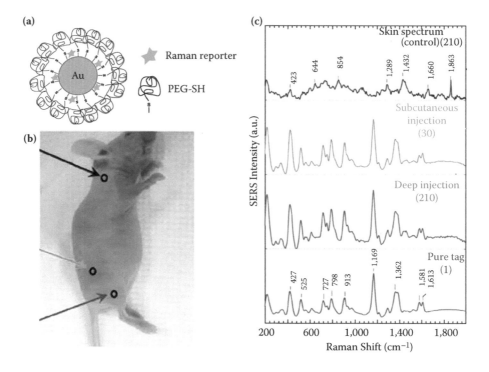

FIGURE 4.7 Example of in vivo surface-enhanced Raman spectroscopy with gold nanoparticles using filter-based planar imaging system. (a) Gold nanoparticles with signal-enhancing reporter could be selectively detected in vivo after subcutaneous and intramuscular injection in mice (b). The SERS spectrum for measurement at injection sites corresponded well to the in vitro spectrum of SERS nanoparticles (bottom), significantly different from tissue without nanoparticles (top). (Adapted for reuse with permission from Qian, X. et al. 2008. *Nature Biotechnology* 26 (1): 83–90.)

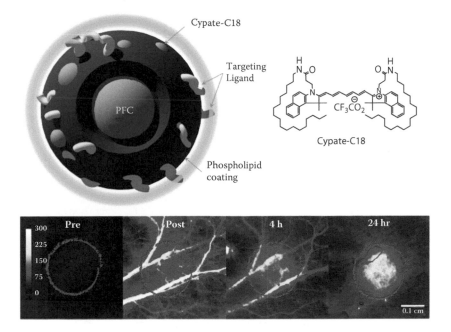

Cypate-C18

Targeting Ligand

PFC

Phospholipid coating

Cypate-C18

CF₃CO₂

| Pre | Post | 4 h | 24 hr |

300
225
150
75
0

0.1 cm

FIGURE 4.9 Perfluorocarbon nanoparticles with perfluorocarbon core (PFC), hydrophobic carbocyanine dye (cypate-C18) in lipid layer, and integrin-targeting ligands extending from the surface. High-resolution fluorescence imaging of integrin expression within the vasculature of intradermal tumors shows nanoparticles selectively accumulate in the neovasculature. (Reused with permission from Akers, W. J. et al. 2010. *Nanomedicine (London)* 5 (5): 715–726.)

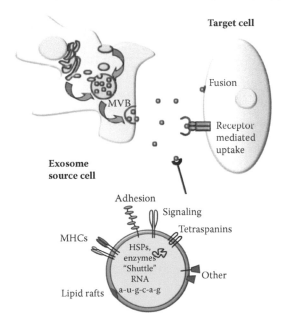

Target cell

Fusion

MVB

Receptor mediated uptake

Exosome source cell

Adhesion

Signaling

Tetraspanins

MHCs

HSPs, enzymes

"Shuttle" RNA

a–u–g–c–a–g

Other

Lipid rafts

FIGURE 4.11 Exosome formation from source cell targeted to another cell. (Used with permission from Hood, J. L. and S. A. Wickline 2012. *Wiley Interdisciplinary Reviews Nanomedicine and Nanobiotechnology* 4 (4): 458–467.)

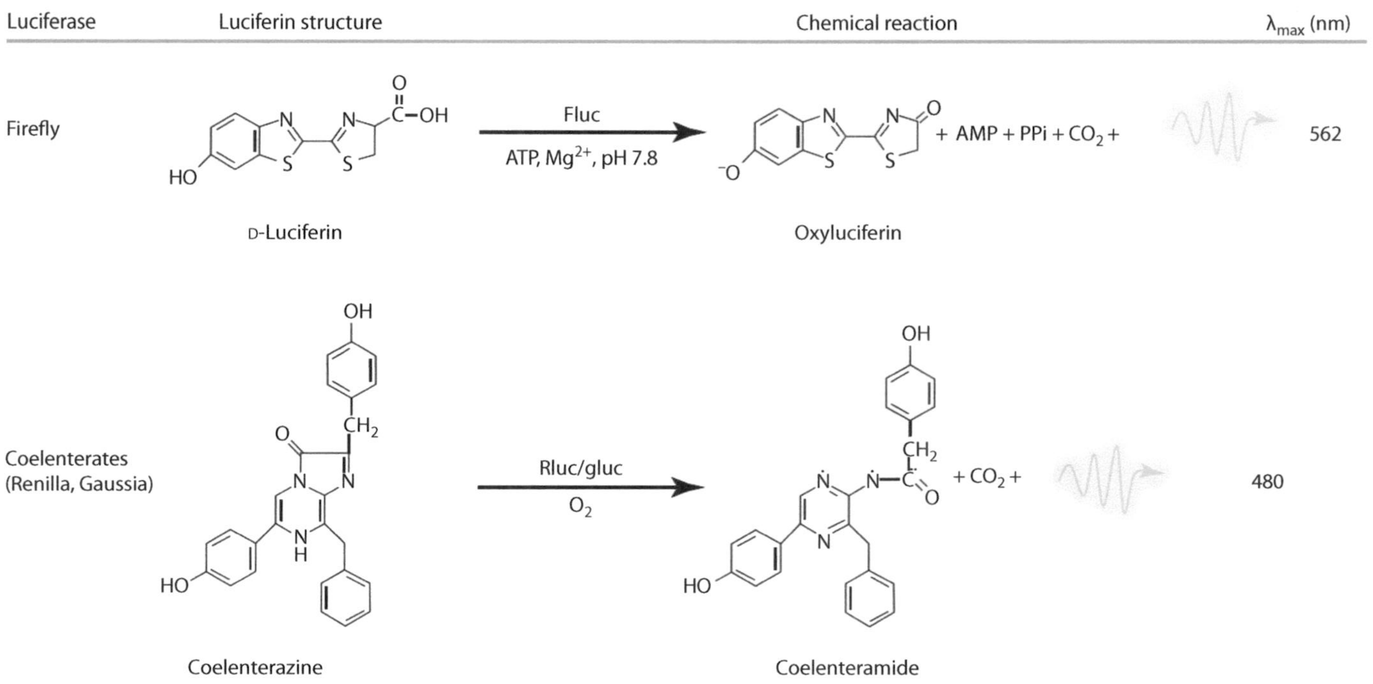

FIGURE 4.17 Chemiluminescent processes that generate light for bioluminescence in cells transfected with firefly (Fluc), Renilla (Rluc), or Gaussia luciferase genes. (Badr, C. E. and B. A. Tannous. 2011. *Trends in Biotechnology* 29 (12): 624–633. With permission.)

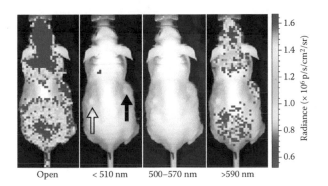

FIGURE 4.19 Imaging of CRET from pseudotumor matrigel implants containing Qtracker705 after intravenous injection of [18F]FDG in a mouse. Cerenkov luminescence was detected throughout the body (open). CRET was detected by employing a 590 nm optical filter for detection of long wavelength emission (>590). (Dothager, R. S. et al. *PLoS One* 5 (10): e13300. With permission.)

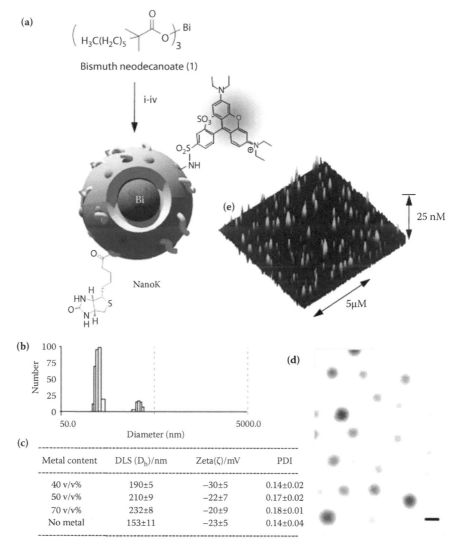

(a)

Bismuth neodecanoate (1)

i-iv

NanoK

(e)

25 nM

5µM

(b)

(d)

(c)

Metal content	DLS (D$_h$)/nm	Zeta(ζ)/mV	PDI
40 v/v%	190±5	−30±5	0.14±0.02
50 v/v%	210±9	−22±7	0.17±0.02
70 v/v%	232±8	−20±9	0.18±0.01
No metal	153±11	−23±5	0.14±0.04

FIGURE 5.1 Synthesis and physicochemical characterization of nano-K. (a) Schematic describing the preparation of bismuth-enriched K-edge nanocolloid (nano-K): (i) suspension of bismuth neodecanoate (1) in sorbitan sesquioleate, vigorously vortex and mix, filter using cotton bed, vortex; (ii) preparation of phospholipid thin film; (iii) resuspension of the thin film in water (0.2 mM); (iv) microfluidization at 4°C, 12,000 psi, 4 min, dialysis (cellulosic membrane, MWCO 20K). (b) Hydrodynamic particle size distribution from DLS; (c) characterization table for three replicates of nano-K; (d) anhydrous state transmission electron microscopy (TEM) images (staining: uranyl acetate; scale bar: 100 nm; (e) atomic force microscope (AFM) image.

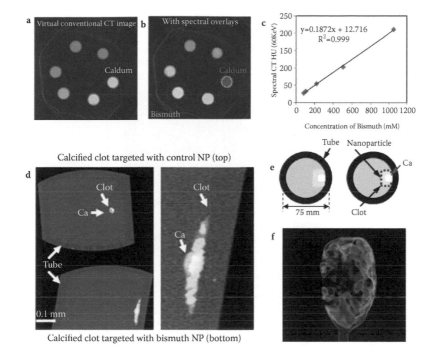

FIGURE 5.2 Spectral CT cross-sectional slices (top) and gradient rendered images (below) of fibrin clots targeted with control (a, e) and nano-K replicates (b–d); integral bismuth distribution in axial slices of fibrin clots: Bound on bismuth layer thickness calculated with scanner spatial resolution at 100 mm, voxel size in reconstructed image (100 mm)³. Bismuth layer thickness: 1 or 2 voxels; bismuth surface density was calculated from integrations perpendicular to the surface layer corresponding to an average 3.5 mass% bismuth for a 100 mm layer thickness.

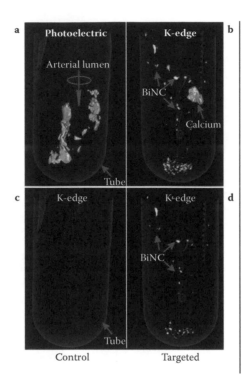

FIGURE 5.3 (a) Photoelectric image of CEA specimen targeted with CoNC showing calcium (presented in red); (b) spectral CT image of CEA specimen targeted with CoNC revealing no presence of the K-edge metal; (c) photoelectric image of CEA specimen targeted with nano-K showing but not differentiating the attenuation contrast of both plaque calcium and the fibrin-targeted nano-K; (d) spectral CT image of CEA specimen after fibrin-targeted nano-K showing the spatial distribution and sparse concentration of thrombus remaining on human carotid specimen. Note: Fibrin-targeted nano-K signal at the bottom of the tube represents thrombus dislodged from CEA during processing.

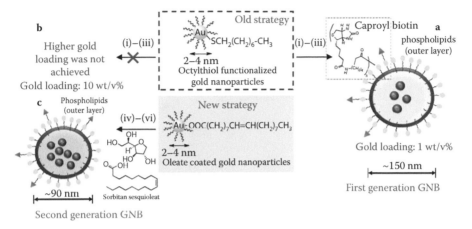

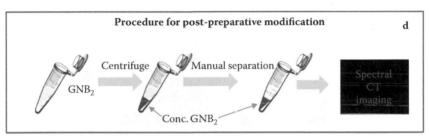

FIGURE 5.5 Synthesis and postpreparative modification of GNB. (a) Previous synthetic approach to first-generation GNB particles from octanethiol-coated AuNP with a final gold loading ca. 2 w/v%: (i) gold nanoparticles suspended in vegetable oil; (ii) preparation of phospholipids thin film; (iii) microfluidization of gold nanoparticle–vegetable oil with surfactants in water. (b) New strategy to increase final gold loading up to 10 w/v%: Oleate-coated gold nanoparticles were suspended in sorbitan sesquioleate and microfluidized with phospholipids thin film mixture as a 2-0-20 formulation. Reaction condition: (i) gold nanoparticles suspended in polysorbate; (ii) preparation of phospholipids thin film; (iii) microfluidization of gold nanoparticle–polysorbate with surfactants in water; 20,000 psi (141 MPa), 4°C, 4 min; (d) schematic representation of the procedure to concentrate GNB2 prior to spectral CT imaging.

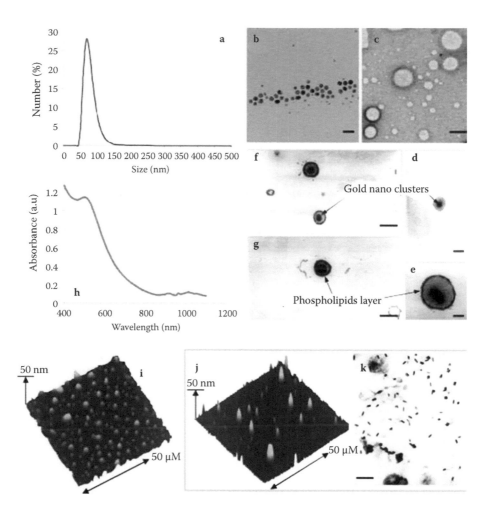

FIGURE 5.6 Characterization of GNB. (a) Number-averaged hydrodynamic diameter of GNB2 from dynamic light scattering measurements; TEM images of (b) oleate-coated gold nanoparticles, scale bar = 20 nm; (c) control nanobeacons with no gold incorporated, scale bar = 100 nm; (d) first-generation GNB (scale bar = 100 nm); (e–g) second-generation GNB revealing the gold clusters entrapped inside the phospholipids membrane (scale bar = 100 nm); (h) UV-vis spectrum of GNB2 in water confirming the presence of gold nanoparticle cluster and their surface plasmon resonance. AFM images of GNB2 (i) and GNB (j, rod); (k) TEM image of GNB (rod) showing the presence of discrete gold rod nanoparticles of larger dimension within a phospholipid encapsulation (scale bar = 100 nm).

4 Optical Imaging with Nanoparticles

Walter J. Akers

CONTENTS

4.1 INTRODUCTION TO OPTICAL IMAGING

Optical contrast for molecular imaging has a rich history from microscopy to endoscopy and fluorescence-guided surgery. Nanomaterials offer significant potential for functions in optical imaging as in other imaging modalities, due to the diverse toolbox available for nanomaterial contrast agent development. Biomedical optical

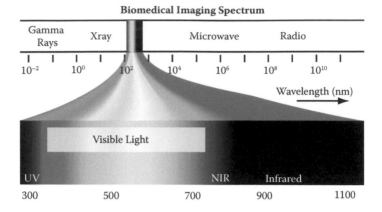

FIGURE 4.1 (**See color insert.**) Comparative energy spectra and location of non-ionizing light in the electromagnetic spectrum.

imaging typically utilizes non-ionizing, visible to near-infrared (NIR) light (450–900 nm) at low radiance levels so that tissues are not damaged. The region of the electromagnetic spectrum occupied by visible light is small, but can be resolved into many "colors" by the human eye and optical detecting instrumentation (Figure 4.1).

Nanomaterials for optical imaging have been developed on many platforms, including organic and inorganic cores with additional shells and coatings to enhance biocompatibility and stability (Figure 4.2). Addition of targeting biomolecules such as peptides and antibodies can enhance tissue-specific delivery for both imaging and therapy. Surface decoration with biocompatible polymers and/or phospholipids is often necessary to prolong circulation time by avoiding natural clearance by the reticuloendothelial system (RES). The choice of nanomaterials for optical imaging is dependent on the biological question of interest as well as the optical characteristics desired. This chapter will discuss the various mechanisms of optical contrast for imaging, and mechanisms for detection and imaging as well as the various nanomaterial contrast agents for biomedical optical imaging.

4.2 OPTICAL IMAGING METHODS

The diffuse nature of light propagation in biological tissues limits penetration depth and spatial resolution in thick samples. Despite these limitations, fluorescence imaging is highly sensitive and offers flexible geometries and many advantages not possible with other modalities (Figure 4.3).

4.2.1 PLANAR REFLECTANCE IMAGING

Camera-based, full-field fluorescence detection is a powerful preclinical imaging technique for fast and economical screening of pharmacokinetics and distribution of fluorescent reporters. Planar reflectance is the simplest and most common geometry for preclinical instrumentation used for fluorescence and bioluminescence imaging

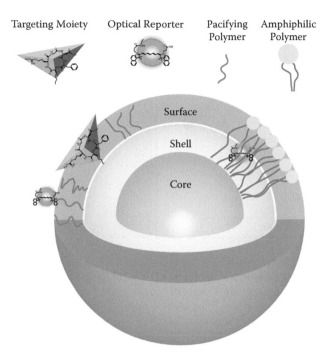

FIGURE 4.2 **(See color insert.)** Simplified cartoon of nanoparticles contrast agent platforms for optical imaging. Nanoparticles are described according to the number of material layers and chemical and biochemical modifications to pacify their surfaces to escape innate immune responses and improve distribution characteristics.

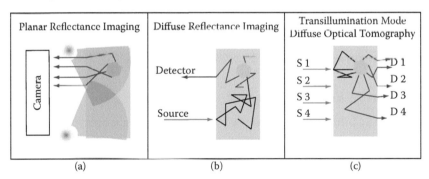

FIGURE 4.3 Optical imaging geometries for fluorescence detection demonstrating (a) planar reflectance, (b) diffuse reflectance, and (c) diffuse transillumination with multiple source (S1–S4) and detector (D1–D4) locations.

(Figure 4.4) [39]. Planar reflectance imaging can provide the highest acquisition speed and resolution for superficial structures but spatial resolution quickly diminishes with depth.

Planar biomedical fluorescence imaging for clinical applications includes fluorescence endoscopy for urologic surgery [60] and robot-assisted laparoscopic surgery

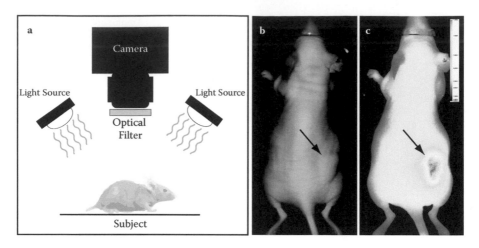

FIGURE 4.4 **(See color insert.)** (a) Example planar reflectance imaging system setup for detection of fluorescence in mouse cancer model. (b) Brightfield and (c) fluorescence images of mouse after intravenous administration of tumor-targeted molecular probe in mouse with subcutaneous tumor (arrow).

[58] as well as fluorescence-guided surgery for brain cancer [44] and ovarian cancer [59]. Acquired two-dimensional images are easy to analyze for high throughput screening of compounds in large study groups. The acquired intensity data are relatively surface weighted as excitation is maximal and attenuation is minimal for fluorescent reporters located at the surface relative to deeper tissues. Another drawback for fluorescence imaging, particularly in the visible wavelength region, is background signal from endogenous fluorophores. Multispectral imaging can be used to separate the signal of interest from these background signals for improved visualization and quantification [59].

4.2.2 Diffuse Reflectance Imaging

Also known as diffuse optical spectroscopy, diffuse reflectance imaging (DRI) utilizes reflectance geometry but with focused excitation and detection of light. This method uses the diffuse nature of light propagation in tissues as a means to extend the depth sensitivity. A comparison of preclinical fluorescence imaging systems demonstrated that DRI gave better contrast than planar reflectance systems for imaging at depths greater than 6 mm [15]. The depth sensitivity of DRI is related to the separation of the excitation source from the detector. This geometry is used in diffuse optical spectroscopy imaging in which a single source-detector pair or multiple source-detector pairs sample the absorption and/or fluorescence over a given region. Mathematical reconstruction of multiple source-detector location measurements into three-dimensional (3D) images (tomography) can also be performed in this geometry [53].

Side   Top

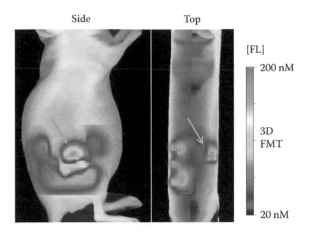

FIGURE 4.5 Fluorescence molecular tomography of mouse with subcutaneous tumor in right flank after injection of IntegriSense 750 NIR fluorescent molecular probe using the FMT2500 (Perkin Elmer). The arrows indicate tumor location. Other areas of fluorescence are background fluorescence from dye in intestines and bladder.

4.2.3 TRANSILLUMINATION IMAGING

Optical imaging in transillumination geometry gives the greatest depth resolution and is the most quantitative method as light passes through the entire tissue. Tomography denotes that mathematical reconstruction algorithms are used to generate 3D whole-body intensity [41] and lifetime [38] maps in preclinical studies using rodent models (Figure 4.5).

Optical mammography using diffuse optical spectroscopy imaging and tomography is being investigated as an alternative or a supplementary technique for mammography for early breast cancer detection [20] and monitoring of therapy [13,42]. At this time, clinical trials of optical mammography have focused on intrinsic contrast from hemoglobin oxygenation and with nonspecific NIR contrast agents ICG and Omocianine [42,47]. Optical mammography has also demonstrated the potential to predict recurrence after chemotherapy by longitudinal changes in tissue optical properties and oxygen saturation [19,52]. Optical mammography will have much greater utility as molecular diagnostic and therapeutic agents, including optically active nanomaterials, become approved for use in humans.

4.2.4 LIGHT PROPAGATION IN BIOLOGICAL TISSUES

Optical contrast for biomedical research and imaging originates from light interaction with tissues and photonic materials. These interactions primarily include photon absorption, scattering, and fluorescence—all wavelength-dependent properties. Unlike high-energy photons used for x-ray and nuclear imaging, which travel in a relatively straight path through tissues, lower energy visible and NIR light travels a more convoluted path. The propagation of light through biological tissues is modeled by the diffusion approximation of the radiative transfer equation (RTE) or more exact computational methods such as iterative Monte Carlo simulations [8,64]:

$$R(\lambda, \rho) = \frac{z_0}{4\pi} \frac{u_s'}{\mu_a + \mu_s'} \left[\left(\mu + \frac{1}{r_1} \right) \frac{3^{-\mu r_1}}{r_1^2} + \left(1 + \frac{4}{3} A \right) \left(\mu + \frac{1}{r_2} \right) \frac{e^{-\mu r_2}}{r_2^2} \right] \quad (4.1)$$

where

$z_0 = (\mu_a + \mu_s')^{-1}$, $\mu = [3\mu_a(\mu_a + \mu_s')]^{1/2}$
$r_1 = (z_0^2 + \rho^2)^{1/2}$
$r_2 = [z_0^2(1 + (4A/3))^2 + \rho^2]^{1/2}$

In Equation (4.1), A is a function of refractive index and μ_a and μ_s are the absorption and scattering coefficients of biological tissues, respectively. Practical use of the RTE for modeling requires assumptions of homogeneity in tissues by using aggregate values for tissue absorption and scattering properties and therefore produces a good approximation of photon transport rather than exact values.

Monte Carlo simulation algorithms numerically calculate the RTE for single "photons" and then add up the total effect to map the probable light distribution in the modeled tissue [64]. This method is very precise and can be used to model unlimited numbers of illumination-detection configurations and tissue layers with various optical properties. The Monte Carlo method is computationally intensive, as millions to billions of photons must be modeled for statistical significance, and is therefore not feasible for application in imaging reconstruction at this time.

4.3　ABSORPTION

Chromophores (light-absorbing molecules) absorb light at specific energies due to resonance of interatomic bonds, in a concentration-dependent manner. The dominant chromophores in biological tissues are oxy- and deoxyhemoglobin, melanin, and water (Figure 4.6). Absorption of light is defined by rearrangement of the Lambert–Beer law:

$$\mu_a(\lambda) = \varepsilon_1 c_1 + \varepsilon_2 c_2 \quad (4.2)$$

where

$\mu_a(\lambda)$ = average distance traveled photon of given wavelength λ before extinction (cm^{-1})
ε = molar extinction coefficient (M^{-1})
c = concentration (M)

This law holds true only for homogeneous, nonscattering solutions in which absorbers do not interact. The concentrations of two components in solution can be determined by measurement of absorbance at two different wavelengths at which ε are known for both components. This method is used to measure the oxygen

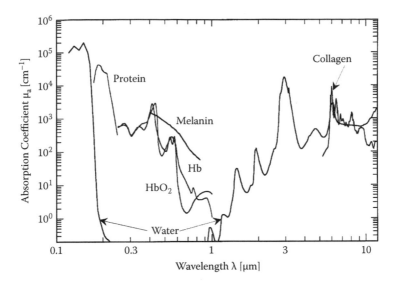

FIGURE 4.6 Absorption spectra of tissues and intrinsic chromophores in biological tissues demonstrating an absorption minima in the near-infrared of about 700–900 nm. (Reproduced with permission from Vogel, A. and Venugopalan, V. 2003. Mechanisms of pulsed laser ablation of biological tissues. *Chem. Rev.* 103 (2): 577–644.)

saturation in tissue by the relative concentrations of oxy- and deoxyhemoglobin in blood.

Wavelength-dependent absorption of light by intrinsic chromophores is responsible for the visible differences diagnostic for many diseases. Contrast agents based on highly absorbing exogenous chromophores can also be useful for diagnostic procedures such as sentinel lymph node biopsy during surgery to remove breast tumors and other cancers. Blue and green dyes such as methylene blue and indocyanine green are routinely used to identify the lymphatic network surrounding the primary breast tumor to diagnose metastatic disease and determine patient prognosis [23,36]. Absorption contrast from exogenous agents generally requires high concentrations, which may be associated with adverse reactions such as allergic reactions and anaphylaxis [36].

4.3.1 PHOTOTHERMAL EFFECTS

Light absorption by nanoparticles using low-energy excitation can be used for photoacoustic imaging. At higher excitation powers and/or prolonged exposure times, nanoparticles can be used to heat tissues for thermal therapy or thermal ablation. Gold nanocages and nanoshells can be more efficient in photothermal ablation than solid gold nanoparticles [49,66]. Raising the temperature within tumor tissue to 40°C–42°C with controlled light irradiation of gold nanoparticles or carbon nanotubes within tissues can eradicate tumors in preclinical models [10]. Selective targeting of nanomaterials to cancer tissues for thermal therapy would provide greater treatment effects while sparing surrounding healthy tissues.

4.4 SCATTERING

Light may also be scattered by particles in a size- and wavelength-dependent manner. Scattering is described as elastic or inelastic scattering. Elastic scattering occurs when the energy of light remains unchanged after interaction. Elastic scattering of light can occur when particles are approximately the same size as the wavelength of light and is therefore directly related to the wavelength of incident light. Light scattering in biological tissue is primarily elastic interaction and is the greatest cause of light attenuation and decrease in spatial resolution for optical imaging methods. The relative scattering coefficients for tissues (μ_s, cm^{-1}) decrease with increasing light wavelength [75].

Raman (inelastic) scattering involves a molecule-specific excitation process in which a change in photon energy occurs. Inelastic scattering of photons resulting in altered frequency of photon. Molecule-dependent events related to chemical bonds—chemical spectroscopy. Contrast agents must have a strong and specific Raman signature; this generally requires high relative concentrations and/or large agents (nanoparticles or microparticles). Coherent anti-Stokes Raman spectroscopy (CARS) is a label-free imaging method while surface-enhanced Raman spectroscopy (SERS) benefits from exongenous contrast agents including plasmonic nanometals [73,74].

4.5 PLASMONIC NANOMETALS

Nanomaterials made from noble metals also have unique optical properties. Metal nanomaterials are not fluorescent, but interact through inelastic Raman scattering. Metal nanomaterials must be coated with pacifying materials to improve biodistribution characteristics and in vivo stability. Fortunately, surface chemistry is relatively easy with metals that contain oxide groups, enabling coating with polymers, silica, biomolecular targeting agents, and signal-enhancing molecules [18].

Plasmon resonance occurs in nanoparticles made from gold, silver, and other metals due to the large radius of electron orbitals confined by the small size of nanoparticles [73]. This phenomenon is not fluorescence, but scattering of light with size-dependent wavelength [25]. Nonspherical particles have different plasmon resonance that is determined by shape [48]. Nanorods have scattering spectra related to the length and diameter of the particles. More complex shapes can have unique plasmon resonance signatures [74].

The SERS signal from plasmonic metal nanoparticles can be enhanced by addition of Raman reporters on the surface [43,70]. This signal enhancement can be greater than six orders of magnitude higher than nanometal alone and enables widefield SERS imaging (Figure 4.7).

Another innovation has been development of gold nanocages that can be loaded with drugs and contrast agents similar to organic nanoparticles (discussed earlier) and also have size-dependent tunable optical properties of gold nanoparticles [65]. The nanocages can be coated with thermoresponsive polymers that form a temperature-sensitive gate mechanism for trapping drugs inside [31]. Upon heating, the polymer gates are opened, releasing the payload at the tissue of interest. The optical properties of gold nanocages can be tuned to NIR absorption for laser-induced heating of the nanoparticles directly for site-directed drug release [66].

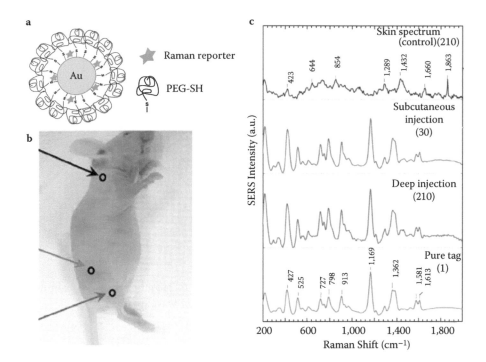

FIGURE 4.7 (See color insert.) Example of in vivo surface-enhanced Raman spectroscopy with gold nanoparticles using filter-based planar imaging system. (a) Gold nanoparticles with signal-enhancing reporter could be selectively detected in vivo after subcutaneous and intramuscular injection in mice (b). The SERS spectrum for measurement at injection sites corresponded well to the in vitro spectrum of SERS nanoparticles (bottom), significantly different from tissue without nanoparticles (top). (Adapted for reuse with permission from Qian, X. et al. 2008. *Nature Biotechnology* 26 (1): 83–90.)

4.6 FLUORESCENCE

Fluorescence is the reemission of light by a molecule after photon absorption and always includes a change in light energy. Fluorescent materials include naturally occurring biomolecules, synthetic organic fluorophores, and some inorganic materials. While intrinsic factors in biological tissues can be fluorescent, they are generally low in concentration and/or quantum yield. Therefore, exogenous fluorescent reporters can produce much greater contrast than can be obtained from absorbing or scattering agents. Exogenous contrast agents can also be synthetically customized to excite and emit at desired wavelengths and high quantum yields for optimal contrast (signal:background). Fluorescent contrast agents have been used for microscopy and in vivo imaging for many years now.

The quantum yield or "brightness" of a fluorescent agent is determined by the product of its absorption coefficient and quantum efficiency as described by

$$Q(\lambda) = \varepsilon * \Phi \qquad (4.3)$$

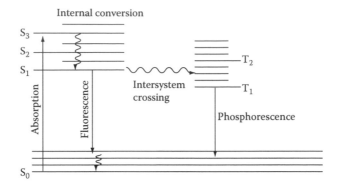

FIGURE 4.8 Energy diagram for photophysical events related to absorption and fluorescence. The fluorescence lifetime is the average time that a population of fluorophores remains in the excited sate (S1–S3) after absorption of photons. (Reused with permission from Berezin, M. Y. and S. Achilefu 2010. *Chemical Reviews* 110 (5): 2641–2684.)

where ε is the absorption coefficient from Equation (4.2) and Φ is the probability of photon emission after absorption at a given excitation wavelength, λ:

$$\Phi = \frac{k_f}{k_f + k_{nr}} \qquad (4.4)$$

where the k_f is the fluorescence rate constant and the nonradiative rate constant (k_{nr}) accounts for all other possibilities dominated by nonradiative decay and heat generation.

As defined before, fluorescence involves the absorption of light followed by emission of a photon of lower energy than the total energy absorbed. The photophysical processes that occur after photon absorption can be described by the energy diagram in Figure 4.8. The process of fluorescence occurs on a picosecond to microsecond timescale. Absorption and fluorescence emission occur almost instantaneously, separated by a brief period in the excited state. The average time that a population of molecules spends in the excited state after photon absorption before fluorescence emission is called the fluorescence lifetime, which is directly related to the quantum yield through Equation (4.4).

For biomedical optical imaging, absorption and fluorescence are the most commonly utilized contrast mechanisms. In general, fluorescence imaging with exogenous contrast agents provides higher sensitivity and better contrast relative to absorption due to the relative amount and spectral characteristics of intrinsic chromophores to fluorophores. Specifically, for deep tissue optical imaging, NIR contrast agents are best suited due to the relative transparency of biological tissues in this wavelength range (700–900 nm) [39].

4.7 ORGANIC NANOMATERIALS FOR FLUORESCENCE IMAGING

Organic fluorophores are abundant and cover the entire spectrum of light from UV to infrared. Organic fluorophores have been customized to optimize quantum yield,

Stokes shift, and excitation/emission wavelengths for many purposes. The photo-physical properties of organic dyes may also be sensitive to environmental factors including pH, temperature, protein binding, solvent effects, etc. Organic dyes may also contain functional groups for easy conjugation to biomolecules and nano-particles. A large field of nanomaterial contrast agents has developed from incor-poration of organic dyes into the nanoparticles structure. The location of the dye in/on the nanoparticles is determined by the production process and the location of functional groups.

4.7.1 MICELLES AND LIPOSOMES

Micelles and liposomes are organic nano- to microsized particles based on amphi-philic molecules that self-assemble in mono- (micelle) or bilayers [32]. Micelles are simple carrier vehicles consisting of amphiphilic phospholipids or block copolymers that form a hydrophobic core when dissolved in an aqueous environment. Micelles are simple to make and good for delivery of hydrophobic drugs, fluorescent report-ers, and hydrophobic nanoparticles [29,67]. Liposomes are surfactant molecules arranged in a bilayer similar to that of a mammalian cell membrane. Liposomes can carry drugs within the hydrophilic core or within the lipid bilayer [32]. Manipulation of the surface characteristics such as adding targeting groups and stealth polymers improves selective delivery of the payload [7,32].

4.7.2 PERFLUOROCARBON

Perfluorocarbons consist of a family of inert fluorine-containing organic compounds initially investigated for their ability to transport oxygen [26]. Perfluorocarbons are not soluble in aqueous or hydrophobic environments. Perfluorocarbon nanoparticles can be formed similarly to micelles, with a perfluorocarbon core surrounded by a hydrophobic and then an amphiphylic layer for aqueous suspensions [26]. Drugs and imaging reporters can be contained in the hydrophobic layer to facilitate delivery of hydrophobic agents, or tethered to the surface as for targeting agents and hydrophilic imaging agents (Figure 4.9) [2,40].

Perfluorocarbon nanoparticles loaded with optical imaging agents have been reported for molecular imaging and targeted drug delivery [1,2,40]. Hydrophobic drugs such as fumagillin, paclitaxel, and others can be sequestered in the hydro-phobic layer [40]. The surface can also be decorated with peptidomimetic or other targeting moieties for site-specific delivery [2]. The perfluorocarbon components are eliminated primarily via respiration after breakdown in the body [26,27].

4.7.3 DENDRIMERS

Polymer science has enabled extraordinary versatility in nanomaterial design. Biocompatible polymer nanoparticles have been designed using nondegradable poly-mers for high in vivo stability, while degradable nanomaterials may be preferable for complete elimination from the body. Polymers may consist of long chains that self-assemble/aggregate or dendrimers that radiate arms for functionalization. Polymers

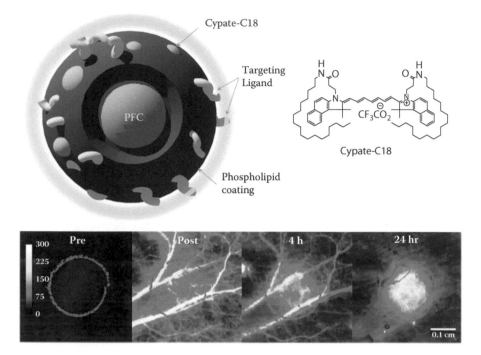

FIGURE 4.9 **(See color insert.)** Perfluorocarbon nanoparticles with perfluorocarbon core (PFC), hydrophobic carbocyanine dye (cypate-C18) in lipid layer, and integrin-targeting ligands extending from the surface. High-resolution fluorescence imaging of integrin expression within the vasculature of intradermal tumors shows nanoparticles selectively accumulate in the neovasculature. (Reused with permission from Akers, W. J. et al. 2010. *Nanomedicine (London)* 5 (5): 715–726.)

such as polyethylene glycol are frequently used as surface coating of nanoparticles to reduce recognition by the reticuloendothelial system (RES) and prolong circulation time. Nanoparticle composition and shape can be formed as desired to optimize in vivo characteristics for circulation, extravasation, and targeted accumulation for imaging and drug delivery.

In general, polymeric nanomaterials are not optically active and must be functionalized with optical reporters for imaging. Thus, polymer nanoparticle design must incorporate functional groups for easy conjugation of organic dyes using systems such as NHS, maleimide, or click chemistry [18].

Fluorescent dendrimer nanoparticles have demonstrated higher photostability and contrast for fluorescence microscopy applications. Kim et al. characterized polyamidoamine (PAMAM) dendrimers with surface-conjugated cyanine dyes with fourfold higher intensity for fluorescence immunoassays [28]. Another novel method was reported to incorporate a fluorophore as the core of a biodegradable dendrimer, protecting the dye from opsonization by proteins (Figure 4.10) [3]. Nanoparticle degradation was reported by changes in fluorescence lifetime of the reporter as it was freed to interact with proteins in tissue. The method and rate of polymer degradation can be customized for biodegradation and clearance from the body.

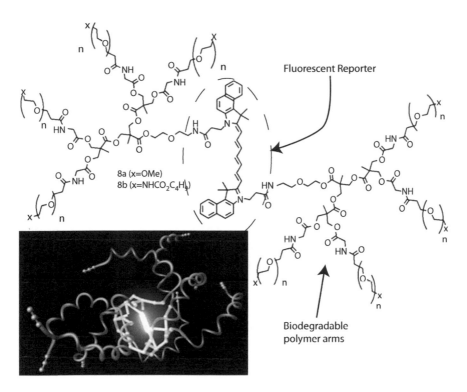

FIGURE 4.10 Biodegradable dendrimer nanoparticle design with NIR fluorescent reporter Cypate as the core. (Reused with permission from Almutairi, A. et al. 2008. *Molecular Pharmacology* 5 (6): 1103–1110.)

4.8 BIOLOGICAL NANOPARTICLES

Naturally occurring nanoparticles including viruses and exosomes are ideally suited for diagnostic and therapeutic applications as they possess intrinsic biological properties for targeted delivery of biological payloads.

4.8.1 VIRUS-BASED NANOPARTICLES

Viruses are natural nanoparticles that carry and specifically deliver DNA and RNA to cells. Fluorescent-labeled viruses have been used to investigate the infection process and have demonstrated promise for biomedical imaging and drug delivery. Delivery of DNA and RNA to cells to alter biological activity could greatly benefit sufferers of many diseases. Naked DNA and RNA are rapidly degraded in the body and therefore cannot be used for therapy. Engineering of viruses for targeted delivery of therapeutic nucleotides is under intense research for gene therapy applications. As such, engineered viral nanoparticles can be labeled with contrast agents for detection of virus distribution and/or contain genetic reporters such as luciferase or fluorescent proteins to report gene expression in target and nontarget cells.

Virus-like [nano]particles (VLPs) include only the viral shell with no genetic material [69]. Viruses that are specific for plants or bacteria but not mammals are ideal choices for these agents as the biological risk is very low. Cowpea mosaic virus (CPMV) and others have been studied intensively for engineering of the exterior and interior [63] for carrying fluorescent reporters and drugs [55]. As these are biologically active agents, they must also be engineered to avoid the immune system for adequate circulation time and tissue distribution.

4.8.2 EXOSOMES

Cells naturally shed smaller vesicle-like subunits for various purposes. Exosomes are nanometer-sized bilayered structures that include membrane-associated proteins as well as cytoplasm (Figure 4.11). Exosomes function as intercellular messengers that can alter the biology of the targeted cells and tissues. Exosomes can carry RNA and other biomolecules, shielding them from degradation by tissue and blood enzymes. The exosomes merge with the cell membranes in target tissues for direct, intracellular delivery of nucleotide signaling molecules. Exosomes have been implicated in diseases including cancer metastasis. Therefore, exosomes are being investigated for diagnosis and therapy as well.

As naturally occurring nanoparticles, exosomes can be harvested and labeled with contrast agents for molecular imaging and theranostic purposes. Exosomes are

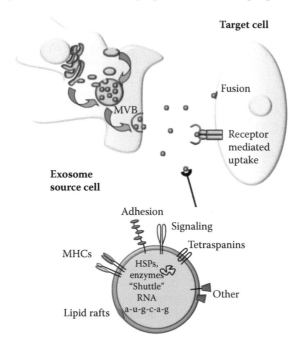

FIGURE 4.11   **(See color insert.)** Exosome formation from source cell targeted to another cell. (Used with permission from Hood, J. L. and S. A. Wickline 2012. *Wiley Interdisciplinary Reviews Nanomedicine and Nanobiotechnology* 4 (4): 458–467.)

constitutively produced by many cell types and can be harvested in vitro or purified from patient blood samples. After purification, exosomes can be labeled with membrane dyes using the same methods for cell labeling for microscopy and in vivo optical imaging. Major obstacles to general use of these bionanoparticles in imaging and therapy involve the routine harvesting of large numbers and adequate characterization without compromising biological activities [24]. As these obstacles are overcome, exosome-based nanoparticles will have a major impact on diagnosis of and therapy for many diseases.

4.9 INORGANIC NANOMATERIALS FOR OPTICAL IMAGING

4.9.1 SILICA-BASED NANOPARTICLES

Silica has unique size-dependent physical and photophysical properties. Silica surface chemistry is well developed, facilitating addition of biomolecules to its surface [18,35]. Silica nanoparticles are most commonly formed using the sol-gel method [62].

Silica nanoparticles that contain well-defined pores, mesoporous silica nanoparticles (MSNs), can be synthesized by modification of the sol-gel method in which silica nanoparticles are formed around surfactant micelles [62]. The porous nature of MSN significantly increases the total surface area for adsorption or conjugation of biomolecules, pharmaceuticals and imaging reporters (Figure 4.12) [46]. Adsorption can be accomplished by simple incubation with hydrophobic dye in anhydrous organic solvent [22]. The mesoporous silica can be degraded inside cells for controlled drug release [6].

Sreejith et al. reported wrapping of squaraine dye-loaded MSN with grapheme oxide sheets for bright and stable fluorescent nanoparticles [54]. Mesoporous silica with embedded fluorescent reporters has also been used as a coating for solid nanoparticles including solid silica [6] and gold nanorods [33].

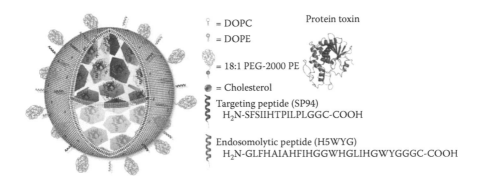

FIGURE 4.12 Cartoon of toxin-loaded, targeted mesoporous silica nanoparticles with covalent and noncovalent loading of interior and exterior with targeting moieties, drugs, and imaging agents. (Reused with permission from Epler et al. 2012.)

4.10   SEMICONDUCTOR NANOCRYSTALS—"QUANTUM DOTS"

Quantum dots (QDs) are nanoparticles composed of semiconductor material that have unique fluorescent properties. The fluorescence properties of these inorganic nanoparticles are governed by quantum physics of semiconductor materials in the nanometer range. Semiconductor materials are not good electrical conductors at room temperature, but electrons can pass from the valence shell to the conducting shell randomly at room temperature and when excited by photon absorption [50]. Transition of an electron from the valence to the conducting shell leaves a "hole" in the valence shell. When these materials are small, the electrons are confined to a small area, but with the same kinetic energy. This is called "quantum confinement." When the electrons return to the valence shell, a photon can be released with energy equal to the bandgap [50].

Therefore, quantum dots have size-tunable, narrow emission spectra (Figure 4.13). Unlike organic fluorophores, quantum dots have broad, energy-dependent absorption spectra with higher extinction coefficients for high-energy light. Because of the broad absorption spectra, quantum dots of various sizes can be excited with the same wavelength, producing multicolored emissions [30]. This characteristic, along with narrow emission spectra, enable simplified multilabel microscopy and imaging. Using quantum dot "bar codes" has also been reported to tag any number of unique objects [21]. Quantum dots also have significantly longer fluorescence lifetimes than organic fluorophores (10–500 ns), enabling time-gated fluorescence imaging for reducing background signals [5].

The semiconductor crystal core of QDs typically contains heavy metals that can be toxic in the body. Quantum dots made from CdSe and CdTe were originally reported in the 1980s and were synthesized in organic solvents. Initial quantum dots were highly hydrophobic. Nie and Alivistros separately overcame this problem, producing water-soluble quantum dots. Since that time, various methods for improved QD synthesis, capping, and decorating have been reported. Much of this work has been done to improve biocompatibility and reduce potential toxicity by protecting the core with a shell of nontoxic material such as safe metal and/or biocompatible polymers [71].

Addition of biocompatible shell can also improve physical qualities of QDs such as those made with PbS, which are very hydrophobic [67,71]. Entrapment of these QDs in micelles improved water solubility while maintaining high quantum yield of NIR emission for in vivo imaging [9]. Other strategies for solubilization include the use of hydrogels, amphiphilic polymers, and thiol substitution [18,67].

Silver-sulfide (Ag2S) and silver-selenium (Ag2Se) nanoparticles have recently been described with fluorescence in the NIR from 700 to 1400 nm [68,72,76]. Preclinical studies have indicated these QDs are not toxic in mice at doses of 15–30 mg/kg [72].

Multicolor imaging with QDs was demonstrated by injecting five different QDs into separate locations for mapping of the lymphatic system in mice [30]. A single excitation source was used to excite all QDs, which were detected by changing the emission filters used during individual acquisitions. In this case, liquid crystal tunable filter was used in the imaging system to capture the emission spectra from the

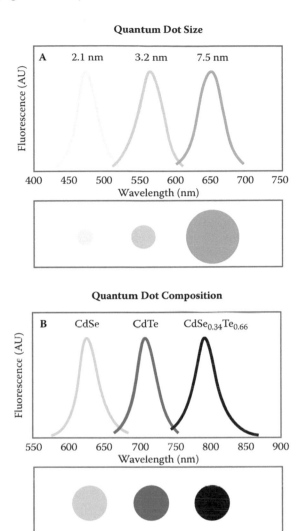

FIGURE 4.13 Size and composition dependence of fluorescence emission from Cd-based QDs. (Smith, A. M. et al. 2004. *Photochemistry and Photobiology* 80 (3): 377–385. With permission.)

single excitation wavelength, which was then separated into distinct images. This study demonstrates the feasibility of multicolor imaging using a single excitation.

Targeted delivery of QDs for molecular imaging has been demonstrated by decoration of QD surfaces with targeting moieties including peptides, protein ligands, and antibodies [12,37]. QDs decorated with RGD peptides Molecular imaging of the angiogenesis marker, $\alpha_v\beta_3$-integrin receptor, in glioblastoma xenografts with -targeted. Molecular imaging in cardiovascular disease using targeted QDs has been reported.

The unique fluorescence properties of QDs also include extended fluorescence lifetimes. The fluorescence lifetime of some QDs extends to 100 ns, significantly longer than autofluorescence and organic fluorescent reporters. It has been shown that fluorescence lifetime imaging can easily separate QD fluorescence from background signals and improve imaging contrast [34].

Many tumors have decreased pH relative to healthy tissues. QDs decorated with pH-sensitive organic fluorophores have recently been designed for pH sensing in vivo. Tang et al. described such a system.

4.11 UPCONVERTING NANOPARTICLES

Upconverting nanoparticles (UCNPs) technology is gaining attention for high-contrast imaging in microscopy and is also making inroads to in vivo diagnostics and therapeutics. Similarly to multiphoton fluorescence imaging, upconverting nanocrystals convert low-energy light to higher energy emission (Figure 4.14).

Water absorption can become an obstacle at wavelengths beyond 900 nm due to sample heating. Zou et al. demonstrated that decoration of the UCNP surface with carbocyanine dyes enabled excitation at 800 nm via fluorescence resonance energy transfer (FRET) with high efficiency [77]. Energy transfer can otherwise be used for imaging and therapy by incorporating photosensitizers onto the surface of UCNP for photodynamic therapy [11,57]. Cui et al. reported folate receptor-targeted NaYF4:Yb,Er UCNP coated with ZnPc photosensitizer for imaging and therapy [14]. UCNPs are much lower in luminescence yield relative to organic dyes and QDs, but have the advantage of lower autofluorescence, similar to multiphoton excitation. Excitation at 980 nm for emission at 700 nm has no autofluorescence background in preclinical imaging relative to 680 nm excitation for the same emission (Figure 4.15) [61].

Upconversion of 980 nm light to 540 nm emission by the nanocrystal core excited the photosensitizer to produce oxygen cytotoxic oxygen radicals (Figure 4.16). Folate receptor-mediated internalization in tumor cells enhanced the cytotoxic effect in vitro and in vivo and enabled tumor-selective fluorescence imaging in tumor-bearing mice. These results demonstrate that the UCNP platform has great potential for imaging and therapy.

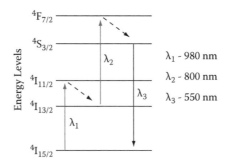

FIGURE 4.14 Energy transition diagram for upconverting nanocrystals. (Berezin, M. Y. and S. Achilefu 2010. *Chemical Reviews* 110 (5): 2641–2684. With permission.)

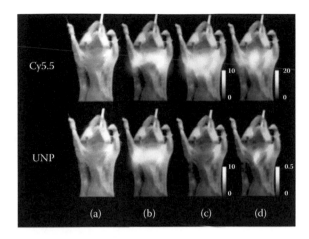

FIGURE 4.15 Fluorescence imaging of mice with tubes containing either NIR dye Cy5.5 (680 nm lex) or erbium- and ytterbium-doped yttrium oxide (Y2O3) upconverting nanoparticles (UCNP, 980 lex) placed into the esophagus. Images shown are brightfield reference images of mouse with (a) demarcated scanning region, (b) intrinsic, (c) autofluorescence, and (d) fluorescence signals from transillumination fluorescence detection at 700 nm overlaid. (Vinegoni, C. et al. 2009. *Optics Letters* 34 (17): 2566–2568. With permission.)

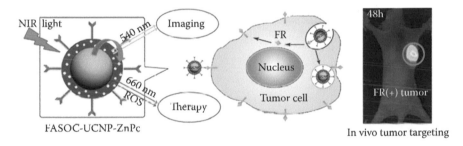

FIGURE 4.16 Cartoon of upconversion process as energy transfer from NaYF4:Yb,Er nanocrystal coated with ZnPc photosensitizer. The UNP core enables excitation at highly tissue-penetrating 980 nm to excite the photosensitizer for imaging and therapy. (Cui, S. et al. 2013. *ACS Nano* 7 (1): 676–688. With permission.)

4.11.1 BIOLUMINESCENCE

Physical and chemical interactions during the excited state can alter the fluorescence yield and/or emission spectra of a fluorophore. These excited state reactions include effects from solvents, temperature, pH, and protein binding as well as interactions with energy-depleting heavy atoms and energy transfer mechanisms such as FRET [44].

Photons can be generated by chemical reactions that change chemical energy into light energy (Figure 4.17). Genetically engineered luminescent reporters most commonly used in biomedical research include enzymes such as firefly and click

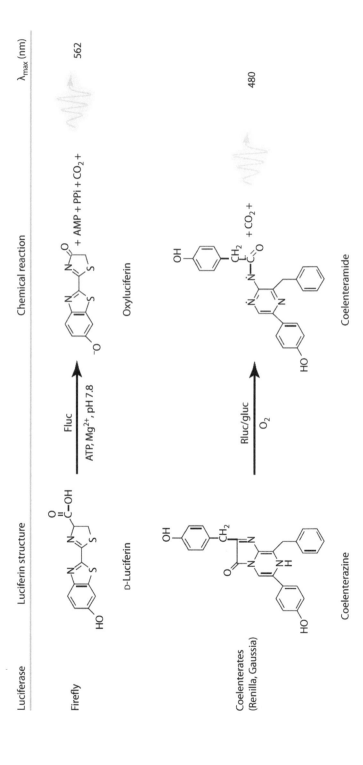

FIGURE 4.17 (**See color insert.**) Chemiluminescent processes that generate light for bioluminescence in cells transfected with firefly (Fluc), Renilla (Rluc), or Gaussia luciferase genes. (Badr, C. E. and B. A. Tannous. 2011. *Trends in Biotechnology* 29 (12): 624–633. With permission.)

beetle luciferases. The genes for photogenic enzymes such as firefly luciferase and others can be inserted into the genome of cells and animals. These enzymes require a separately administered substrate such as luciferin or coelenterazine to produce luminescence in an energy-dependent process.

Bioluminescence imaging (BLI) has been used to monitor gene and related protein expression, protein–protein interactions, and track bacterial and mammalian cells in animal models [4]. BLI requires very sensitive cameras with low noise for detection. Due to visible wavelength emission, BLI is limited in depth of detection and resolution is highly depth dependent. Despite emission at visible wavelengths, biolumines-cence can produce very high contrast because of almost zero intrinsic luminescence from mammalian tissues. BLI is widespread in basic biomedical research.

To shift the emission of BLI to longer wavelengths for better depth resolution, researchers have worked to discover longer wavelength photoenzymes [4]. Another strategy has been to conjugate luciferase enzyme to the surface of quantum dots for bioluminescence resonance energy transfer (BRET) with tunable emission into the NIR. While the photon yield of this process is low, the optimized emission wave-length enables deeper signal detection and higher resolution relative to the natu-ral visible emission (Figure 4.18) [51]. Dragavon et al. demonstrated that significant energy can be transferred via radiative processes from luminescent bacteria to quan-tum dots over distances greater than 10 nm for red-shifting of bioluminescence [17]. The difficulty in development of BRET-based imaging agents is the necessity for conjugating luciferase enzymes to quantum dots while retaining enzyme activity.

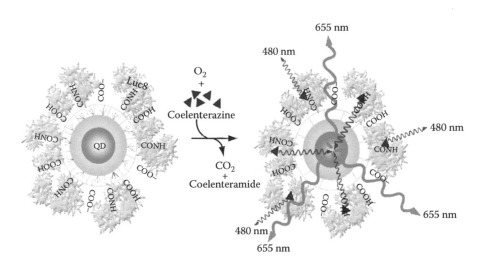

FIGURE 4.18 Cartoon of luciferase-decorated QD for bioluminescence resonance energy transfer resulting in emission at far red wavelengths. (So, M.-K. et al. 2006. *Nature Biotechnology* 24 (3): 339–343. With permission.)

4.11.2 CHEMILUMINESCENCE

Luminescence can also be generated during the decay of high-energy radioactive decay in a process called Cerenkov luminescence. Recently, Cerenkov emission imaging has gained interest for biomedical imaging using compact and economical high-sensitivity cameras as used for bioluminescence detection. Cerenkov emission occurs when energy particles travel faster than the speed of light, which can occur in media with refractive indexes greater than that of a vacuum, such as water and biological tissues [45]. High-energy photons such as positrons are required to overcome the threshold for CE generation [45].

The wavelength of light produced is related to the velocity of the particle, which slows over time; therefore, CE is most intense in the UV and blue and decreases with increasing wavelength. Due to the poor transmission of CE through tissues, researchers have investigated methods to transfer CE energy to reporters that then emit at longer wavelengths. As with BRET, the most successful of these has been with energy transfer to QDs as they excite best in UV and can have tuned emission to the NIR. In vitro chemiluminescence resonance energy transfer (CRET) has been demonstrated by mixing radionuclides with QDs and other optically active nanomaterials. In vivo CRET imaging was reported by Dothater et al. where long wavelength emission from QD-containing pseudotumors was detected following injection of positron-emitting radionuclide (Figure 4.19) [16]. CE and CRET imaging are unique new applications of optical imaging that can be useful for high-throughput screening of radiotracer uptake and potential clinical applications [56].

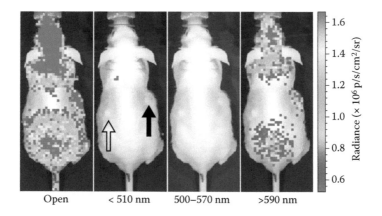

Open   < 510 nm   500–570 nm   >590 nm

FIGURE 4.19 **(See color insert.)** Imaging of CRET from pseudotumor matrigel implants containing Qtracker705 after intravenous injection of [18F]FDG in a mouse. Cerenkov luminescence was detected throughout the body (open). CRET was detected by employing a 590 nm optical filter for detection of long wavelength emission (>590). (Dothager, R. S. et al. *PLoS One* 5 (10): e13300. With permission.)

4.12 SUMMARY

Biomedical optical imaging has many applications in clinical diagnostics and will continue to grow in use as diagnostic and theranostic nanoparticles achieve FDA approval. A primary benefit of optical imaging includes the great dynamic range of resolution from nanoscopy to deep tissue imaging. The versatility of optical detection techniques and variety of optical contrast mechanisms demonstrates a broad potential for further development of in vitro and in vivo diagnostic applications. Optical imaging is limited in depth resolution, preventing whole-body imaging, but provides complementary information to other radiologic modalities and can provide real-time feedback, which is not possible with other modalities.

Nanomaterial development for optical imaging continues to expand rapidly due to advances in photonic materials, spectroscopy, and biochemistry. These advances propel development of new imaging agents, which then drive further scientific discoveries. Nanomaterials represent a vast toolbox of multifunctional agents for delivery of drugs, biological modifiers, and diagnostic agents. Optically active nanoparticles are important parts of this toolbox for development of theranostic agents by enabling in vivo pharmacokinetic analysis and histologic examination of biodistribution without harmful radiation or damage to tissues. Construction of nanoparticles for optical imaging has become relatively straightforward even as nanoparticle design has become increasingly complex.

Fluorescent nanoparticles utilizing organic fluorescent reporters and inorganic semiconductor fluorescent materials have established a significant presence in preclinical molecular imaging and drug delivery development. Challenges remain in development and translation of nanoparticles for biomedical optical imaging. Few optical imaging agents have been approved for use in humans for incorporation into nanoparticles. Fine-tuning of delivery into targeted tissues and cells is needed for more specific contrast and selective detection of molecular events. The great advances in the field of optical nanoparticles demonstrate the potential for continued growth in this field and improved patient outcomes in the near future.

REFERENCES

1. Akers, W. J., C. Kim, et al. 2011. Noninvasive photoacoustic and fluorescence sentinel lymph node identification using dye-loaded perfluorocarbon nanoparticles. *ACS Nano* 5 (1): 173–182.
2. Akers, W. J., Z. Zhang, et al. 2010. Targeting of alpha(nu)beta(3)-integrins expressed on tumor tissue and neovasculature using fluorescent small molecules and nanoparticles. *Nanomedicine (London)* 5 (5): 715–726.
3. Almutairi, A., W. J. Akers, et al. 2008. Monitoring the biodegradation of dendritic near-infrared nanoprobes by in vivo fluorescence imaging. *Molecular Pharmaceutics* 5 (6): 1103–1110.
4. Badr, C. E. and B. A. Tannous 2011. Bioluminescence imaging: Progress and applications. *Trends in Biotechnology* 29 (12): 624–633.
5. Berezin, M. Y. and S. Achilefu 2010. Fluorescence lifetime measurements and biological imaging. *Chemical Reviews* 110 (5): 2641–2684.

6. Bergman, L., P. Kankaanpaa, et al. 2013. Intracellular degradation of multilabeled poly(ethylene imine)-mesoporous silica-silica nanoparticles: Implications for drug release. *Molecular Pharmaceutics* 10:1795–1803.

7. Blanco, E., A. Hsiao, et al. 2011. Nanomedicine in cancer therapy: Innovative trends and prospects. *Cancer Science* 102 (7): 1247–1252.

8. Boas, D. A., C. Pitris, et al. 2011. *Handbook of biomedical optics.* Boca Raton, FL: CRC Press.

9. Cao, J., H. Zhu, et al. 2012. In vivo NIR imaging with PbS quantum dots entrapped in biodegradable micelles. *Journal of Biomedical Materials Research A* 100 (4): 958–968.

10. Chatterjee, D. K., P. Diagaradjane, et al. 2011. Nanoparticle-mediated hyperthermia in cancer therapy. *Therapy Delivery* 2 (8): 1001–1014.

11. Chatterjee, D. K. and Z. Yong. 2008. Upconverting nanoparticles as nanotransducers for photodynamic therapy in cancer cells. *Nanomedicine (London)* 3 (1): 73–82.

12. Cheng, K. and Z. Cheng 2012. Near infrared receptor-targeted nanoprobes for early diagnosis of cancers. *Current Medicinal Chemistry* 19 (28): 4767–4785.

13. Collettini, F., J. C. Martin, et al. 2012. Diagnostic performance of a near-infrared breast imaging system as adjunct to mammography versus X-ray mammography alone. *European Radiology* 22 (2): 350–357.

14. Cui, S., D. Yin, et al. 2013. In vivo targeted deep-tissue photodynamic therapy based on near-infrared light triggered upconversion nanoconstruct. *ACS Nano* 7 (1): 676–688.

15. de la Zerda, A., S. Bodapati, et al. 2010. A comparison between time domain and spectral imaging systems for imaging quantum dots in small living animals. *Molecular Imaging Biology* 12 (5): 500–508.

16. Dothager, R. S., R. J. Goiffon, et al. 2010. Cerenkov radiation energy transfer (CRET) imaging: A novel method for optical imaging of PET isotopes in biological systems. *PLoS One* 5 (10): e13300.

17. Dragavon, J., S. Blazquez, et al. 2012. In vivo excitation of nanoparticles using luminescent bacteria. *Proceedings of National Academy of Sciences USA* 109 (23): 8890–8895.

18. Erathodiyil, N. and J. Y. Ying 2011. Functionalization of inorganic nanoparticles for bioimaging applications. *Accounts of Chemical Research* 44 (10): 925–935.

19. Falou, O., H. Soliman, et al. 2012. Diffuse optical spectroscopy evaluation of treatment response in women with locally advanced breast cancer receiving neoadjuvant chemotherapy. *Translational Oncology* 5 (4): 238–246.

20. Fang, Q., J. Selb, et al. 2011. Combined optical and x-ray tomosynthesis breast imaging. *Radiology* 258 (1): 89–97.

21. Fournier-Bidoz, S., T. L. Jennings, et al. 2008. Facile and rapid one-step mass preparation of quantum-dot barcodes. *Angewandte Chemie* International Edition England 47 (30): 5577–5581.

22. Gianotti, E., C. A. Bertolino, et al. 2009. Photoactive hybrid nanomaterials: Indocyanine immobilized in mesoporous MCM-41 for in-cell bioimaging. *ACS Applied Material Interfaces* 1 (3): 678–687.

23. Gipponi, M. 2005. Clinical applications of sentinel lymph-node biopsy for the staging and treatment of solid neoplasms. *Minerva Chirurgica* 60 (4): 217–233.

24. Hood, J. L. and S. A. Wickline 2012. A systematic approach to exosome-based translational nanomedicine. *Wiley Interdisciplanary Reviews Nanomedicine and Nanobiotechnology* 4 (4): 458–467.

25. Huang, X., P. K. Jain, et al. 2007. Gold nanoparticles: Interesting optical properties and recent applications in cancer diagnostics and therapy. *Nanomedicine (London)* 2 (5): 681–693.

26. Janjic, J. M. and E. T. Ahrens 2009. Fluorine-containing nanoemulsions for MRI cell tracking. *Wiley Interdisciplinary Reviews of Nanomedicine and Nanobiotechnology* 1 (5): 492–501.

27. Kaneda, M. M., S. Caruthers, et al. 2009. Perfluorocarbon nanoemulsions for quantitative molecular imaging and targeted therapeutics. *Annals of Biomedical Engineering* 37 (10): 1922–1933.

28. Kim, Y., S. H. Kim, et al. 2013. Dendrimer probes for enhanced photostability and localization in fluorescence imaging. *Biophysics Journal* 104 (7): 1566–1575.

29. Koole, R., W. J. Mulder, et al. 2009. Magnetic quantum dots for multimodal imaging. *Wiley Interdisciplinary Reviews of Nanomedicine and Nanobiotechnology* 1 (5): 475–491.

30. Kosaka, N., T. E. McCann, et al. 2010. Real-time optical imaging using quantum dot and related nanocrystals. *Nanomedicine (London)* 5 (5): 765–776.

31. Li, W., X. Cai, et al. 2011. Gold nanocages covered with thermally-responsive polymers for controlled release by high-intensity focused ultrasound. *Nanoscale* 3 (4): 1724–1730.

32. Lopez-Davila, V., A. M. Seifalian, et al. 2012. Organic nanocarriers for cancer drug delivery. *Current Opinions in Pharmacology* 12 (4): 414–419.

33. Luo, T., P. Huang, et al. 2011. Mesoporous silica-coated gold nanorods with embedded indocyanine green for dual mode x-ray CT and NIR fluorescence imaging. *Optics Express* 19 (18): 17030–17039.

34. Ma, G. 2013. Background-free in vivo time domain optical molecular imaging using colloidal quantum dots. *ACS Applied Material Interfaces* 5:2835–2844.

35. Mader, H., X. Li, et al. 2008. Fluorescent silica nanoparticles. *Annals of New York Academy of Sciences* 1130: 218–223.

36. Masannat, Y., H. Shenoy, et al. 2006. Properties and characteristics of the dyes injected to assist axillary sentinel node localization in breast surgery. *European Journal of Surgical Oncology* 32 (4): 381–384.

37. Michalet, X., F. F. Pinaud, et al. 2005. Quantum dots for live cells, in vivo imaging, and diagnostics. *Science* 307 (5709): 538–544.

38. Nothdurft, R. E., S. V. Patwardhan, et al. 2009. In vivo fluorescence lifetime tomography. *Journal of Biomedical Optics* 14 (2): 024004.

39. Ntziachristos, V., J. Ripoll, et al. 2005. Looking and listening to light: The evolution of whole-body photonic imaging. *Nature Biotechnology* 23 (3): 313–320.

40. O'Hanlon, C. E., K. G. Amede, et al. 2012. NIR-labeled perfluoropolyether nanoemulsions for drug delivery and imaging. *Journal of Fluorescent Chemistry* 137: 27–33.

41. Patwardhan, S. V., S. Bloch, et al. 2006. Quantitative small animal fluorescence tomography using an ultrafast gated image intensifier. *Conference Proceedings of IEEE Engineering in Medicine and Biology Society* 1: 2675–2678.

42. Poellinger, A., T. Persigehl, et al. 2011. Near-infrared imaging of the breast using omocianine as a fluorescent dye: Results of a placebo-controlled, clinical, multicenter trial. *Investigative Radiology* 46 (11): 697–704.

43. Qian, X., X. H. Peng, et al. 2008. In vivo tumor targeting and spectroscopic detection with surface-enhanced Raman nanoparticle tags. *Nature Biotechnology* 26 (1): 83–90.

44. Roberts, D. W., P. A. Valdes, et al. 2012. Glioblastoma multiforme treatment with clinical trials for surgical resection (aminolevulinic acid). *Neurosurgery Clinics of North America* 23 (3): 371–377.

45. Robertson, R., M. S. Germanos, et al. 2009. Optical imaging of Cerenkov light generation from positron-emitting radiotracers. *Physics in Medicine and Biology* 54 (16): N355–365.

46. Rosenholm, J. M., C. Sahlgren, et al. 2010. Towards multifunctional, targeted drug delivery systems using mesoporous silica nanoparticles—Opportunities & challenges. *Nanoscale* 2 (10): 1870–1883.

47. Schneider, P., S. Piper, et al. 2011. Fast 3D near-infrared breast imaging using indocyanine green for detection and characterization of breast lesions. *Rofo* 183 (10): 956–963.

48. Seekell, K., H. Price, et al. 2012. Optimization of immunolabeled plasmonic nanoparticles for cell surface receptor analysis. *Methods* 56 (2): 310–316.
49. Shen, H., J. You, et al. 2012. Cooperative, nanoparticle-enabled thermal therapy of breast cancer. *Advances in Healthcare Materials* 1 (1): 84–89.
50. Smith, A. M., X. Gao, et al. 2004. Quantum dot nanocrystals for in vivo molecular and cellular imaging. *Photochemistry and Photobiology* 80 (3): 377–385.
51. So, M.-K., C. Xu, et al. 2006. Self-illuminating quantum dot conjugates for in vivo imaging. *Nature Biotechnology* 24 (3): 339–343.
52. Soliman, H., A. Gunasekara, et al. 2010. Functional imaging using diffuse optical spectroscopy of neoadjuvant chemotherapy response in women with locally advanced breast cancer. *Clinical Cancer Research* 16 (9): 2605–2614.
53. Solomon, M., B. R. White, et al. 2011. Video-rate fluorescence diffuse optical tomography for in vivo sentinel lymph node imaging. *Biomedical Optics Express* 2 (12): 3267–3277.
54. Sreejith, S., X. Ma, et al. 2012. Graphene oxide wrapping on squaraine-loaded mesoporous silica nanoparticles for bioimaging. *Journal of American Chemical Society* 134 (42): 17346–17349.
55. Steinmetz, N. F., A. L. Ablack, et al. 2011. Intravital imaging of human prostate cancer using viral nanoparticles targeted to gastrin-releasing peptide receptors. *Small* 7 (12): 1664–1672.
56. Thorek, D., R. Robertson, et al. 2012. Cerenkov imaging—A new modality for molecular imaging. *American Journal of Nuclear Medicine and Molcular Imaging* 2 (2): 163–173.
57. Tian, G., W. Ren, et al. 2012. Red-emitting upconverting nanoparticles for photodynamic therapy in cancer cells under near-infrared excitation. *Small* 9:1929–1938.
58. Tobis, S., J. K. Knopf, et al. 2012. Robot-assisted and laparoscopic partial nephrectomy with near infrared fluorescence imaging. *Journal of Endourology* 26 (7): 797–802.
59. van Dam, G. M., G. Themelis, et al. 2011. Intraoperative tumor-specific fluorescence imaging in ovarian cancer by folate receptor-alpha targeting: First in-human results. *Nature Medicine* 17 (10): 1315–1319.
60. van den Berg, N. S., F. W. van Leeuwen, et al. 2012. Fluorescence guidance in urologic surgery. *Current Opinions in Urology* 22 (2): 109–120.
61. Vinegoni, C., D. Razansky, et al. 2009. Transillumination fluorescence imaging in mice using biocompatible upconverting nanoparticles. *Optics Letters* 34 (17): 2566–2568.
62. Vivero-Escoto, J. L., R. C. Huxford-Phillips, et al. 2012. Silica-based nanoprobes for biomedical imaging and theranostic applications. *Chemical Society Reviews* 41 (7): 2673–2685.
63. Wen, A. M., S. Shukla, et al. 2012. Interior engineering of a viral nanoparticle and its tumor homing properties. *Biomacromolecules* 13 (12): 3990–4001.
64. Wilson, R. H. and M. A. Mycek 2011. Models of light propagation in human tissue applied to cancer diagnostics. *Technology in Cancer Research Treatment* 10 (2): 121–134.
65. Xia, X., M. Yang, et al. 2011. An enzyme-sensitive probe for photoacoustic imaging and fluorescence detection of protease activity. *Nanoscale* 3 (3): 950–953.
66. Xia, Y., W. Li, et al. 2011. Gold nanocages: From synthesis to theranostic applications. *Accounts of Chemical Research* 44 (10): 914–924.
67. Xue, B., J. Cao, et al. 2012. Four strategies for water transfer of oil-soluble near-infrared-emitting PbS quantum dots. *Journal of Materials Science Materials in Medicine* 23 (3): 723–732.
68. Yang, H. Y., Y. W. Zhao, et al. 2013. One-pot synthesis of water-dispersible Ag2S quantum dots with bright fluorescent emission in the second near-infrared window. *Nanotechnology* 24 (5): 055706.

69. Yildiz, I., S. Shukla, et al. 2011. Applications of viral nanoparticles in medicine. *Current Opinions in Biotechnology* 22 (6): 901–908.
70. Zavaleta, C. L., B. R. Smith, et al. 2009. Multiplexed imaging of surface enhanced Raman scattering nanotags in living mice using noninvasive Raman spectroscopy. *Proceedings of National Academy of Sciences USA* 106 (32): 13511–13516.
71. Zhang, F., E. Lees, et al. 2011. Polymer-coated nanoparticles: A universal tool for bio-labelling experiments. *Small* 7 (22): 3113–3127.
72. Zhang, Y., G. Hong, et al. 2013. Biodistribution, pharmacokinetics and toxicology of Ag2S near-infrared quantum dots in mice. *Biomaterials* 34 (14): 3639–3646.
73. Zhang, Y., H. Hong, et al. 2010. Imaging with Raman spectroscopy. *Current Pharmaceutical Biotechnology* 11 (6): 654–661.
74. Zheng, Y. B., B. Kiraly, et al. 2012. Molecular plasmonics for biology and nanomedicine. *Nanomedicine (London)* 7 (5): 751–770.
75. Zhu, C., G. M. Palmer, et al. 2008. Diagnosis of breast cancer using fluorescence and diffuse reflectance spectroscopy: A Monte-Carlo-model-based approach. *Journal of Biomedical Optics* 13 (3): 034015.
76. Zhu, C. N., P. Jiang, et al. 2013. Ag(2)Se quantum dots with tunable emission in the second near-infrared window. *ACS Applied Material Interfaces* 5 (4): 1186–1189.
77. Zou, W. Q., C. Visser, et al. 2012. Broadband dye-sensitized upconversion of near-infrared light. *Nature Photonics* 6 (8): 560–564.

5 Contrast Agents for Computed Tomographic Imaging

Dipanjan Pan, Santosh K. Misra, and Sumin Kim

CONTENTS

5.1 INTRODUCTION

Medical imaging infers detailed pictures of processes inside the subject's body without an invasive approach. In a nearly decade-old development, molecular imaging has gained attention as a noninvasive approach to cellular and subcellular imaging explaining chemical and biological processes. Chemistry, biology, and engineering and their triad combinations have led from clinical imaging to biochemically based assessments surging from a mere anatomical static picture of the situation. Such developments would provide information that is unattainable with other imaging technologies as it identifies disease or a functional or building block abnormality in its earliest stages and determines its exact location—often before symptoms occur or abnormalities can be detected with other diagnostic tests [1–4].

Computed tomography (CT), magnetic resonance (MR) imaging, and ultrasonography (US) are general clinical anatomical imaging modalities used nowadays [5]. CT has the capacity to tomographically scan large portions of a body quickly (unlike other techniques), but its soft-tissue contrast is not reported to be as good as required. Hence, a targeted molecular imaging with CT might show a relatively insensitive nature to probe detection compared to the other modalities, especially nuclear imaging. Newer approaches based on nanoparticle chemistries showed very high intrinsic metal density in order to achieve x-ray detectable concentrations in a CT imaging approach [6]. To assist this benefit, improved chemistry has been supportive of rapid image acquisition and statistical iterative reconstruction algorithms. These algorithmic advantages could be reflected as a reduction in the imaging noise floor, making it a more sensitive contrast detection than those obtained with traditional back project reconstruction routines [7,8].

Generic development of CT instrumentation has achieved new heights with expansion of simultaneous slice numbers from 1 up to 320 simultaneous slices to introduction of kVp switching, dual-energy detectors, the preclinical emergence of phase-contrast x-ray imaging, and fluorescent CT imaging. A special interest has been evolved other than SPECT/CT (a hybrid single-photon emission computed tomography and computed tomography instrument) by the introduction of spectral CT imaging based on K-edge detection [9,10]. In a newer advancement, spectral or multicolor CT can acquire a traditional CT image with detection of certain elements within the x-ray bandwidth based on their distinctive K-edge energies [11]. An emergent increase in the attenuation coefficient of photons falling out at an energy exceeding the binding energy of the K shell electron of the atoms within the x-ray beam defines the K-edge energy of an element.

High accumulations of directed elements attached to pathologic biomarkers may be presented on the typical CT image as colorized voxels during K-edge imaging. These slots of high attenuation voxels are promptly differentiated from other distinctly lower K-edge energies attenuating sources, such as calcium deposits. Furthermore, photon counts can reveal the quantization of regions of attenuation by K-edge metals. The photon-counting detectors fetch spectral information across the bandwidth of the emitted photons, whereas conventional CT furnishes information about the overall attenuation of the emitted x-ray beam. In urgent need of clinical cases, CT molecular imaging of intracoronary microthrombus associated with ruptured plaque is required to differentiate potential false-positive contrast signals derived from intraplaque calcium deposits so that patients admitted with chest pain can be properly diagnosed [8–11].

5.2 CONTRAST AGENTS: THE FUNDAMENTALS

High-atomic-number (Z) elements that absorb x-rays constitute contrast materials of particular interest in the field of advanced CT imaging. Among these elements, iodine, a blood pool CT contrast agent, induces additional K-edge-based attenuation in the compound. The rich abundance of photons within an x-ray beam of appropriate energy in case of iodine is partially achieved by its high atomic number. A very strong inherent attenuation with a K-edge energy value near the top of the x-ray beam bandwidth can be correlated to the very high atomic number of the white insoluble powder barium (barium sulfate). The flaws in terms of K-edge imaging associated with iodinated contrast agents make them nearly unusable in clinics because of photon starvation. This photon starvation generates due to the impact of x-ray beam hardening (absorption of lower energy photons), relatively low K-edge energy ($I = 33$ keV; $Ba = 37$ keV), and omnidirectional x-ray scattering. These drawbacks of widely used iodinated contrast agents originated interest in the development of smartly engineered nanomaterials ("nanosmarts") as CT contrast agents encompassing other high atomic number metals—namely, gold (Au), gadolinium (Gd), bismuth (Bi), and tantalum (Ta) [12]. Thus, specific CT molecular imaging applications are on top of multitudinous efforts to generate a suitable smartly engineered nanomaterial. The inherent characteristic of such nanosmarts must be excellent radio opacity, easy and cost-effective synthetic origin, sincere biocompatibility, purposeful in vivo stability (i.e., particularly during circulatory transit to targets and through the imaging period), and physical and chemical stability with strong tolerance to sterilization to make a better one.

The chelated metal complexes (Gd-DTPA or Yb-DTPA) [13,14] may be small molecular weight compounds with the identity of blood pool contrast agents, but functionalization with homing ligands may not bring adequate metal for site-specific CT detection. It urges the high mass delivery of high atomic weight elements for successful molecular imaging for CT, where high atomic weight elements might be packaged into a nanoscale particle as a "nanocontrast" agent. These nanocontrast agents may be categorized based on metal involved as gadolinium nanoparticles, gold nanoparticles with different size and morphology (spherical vs. rod, cages), tantalum-encapsulated, and ytterbium-based agents.

5.3 METAL NANOPARTICLES

Evolving "nanocontrast" agents with virtue of CT imaging has been quite a challenge. Materials with very high metal densities and radio opacity are urgently needed because of the insensitivity of x-ray-based techniques. Also, these materials must be easy to store for a longer time before use but should not lose their property in the blood circulation for hours after being administered as an i.v. These materials are required not to be metabolized or transmetallated in vivo, and the metal must be completely cleared from the subject's body. The "hard" crystalline metal nanoparticles have been a fruitful task by nanomaterial scientists [12]. The combination of satisfactory physicochemical properties makes them important players for CT imaging. High atomic number metals such as gold, bismuth, gadolinium, and tantalum offer stronger x-ray attenuation and much CT contrast compared to iodine and barium [6,12]. To take further advantage of the metal nanoparticles, they can be functionalized with biologically relevant molecules for targeting or drug delivery because of their high affinity toward reactive groups (e.g., thiol, disulfide amines, etc.) [1].

Metallic nanocrystal-based nanocontrast agents are well studied now [6,12]. Although resistant to biometabolism, the nanocontrast agents based on heavy-metal crystals (oxides, sulfides, etc.) are not biologically eliminated. The size of core crystals determines the fate across in vivo biodistribution, pharmacokinetics, and bio-elimination. Such nanocrystals with possibly >6 nm avoid the renal elimination to enhance their biocirculation life span undesirably in subjects' bodies. To add to the challenge, hepatic elimination of such "hard" particles through bile is prevented, except in rodents with permissively open biliary canaliculi, which may diminish the possibility of their long-term safety.

In a recent advancement, colloidal "soft" nanoparticles with the ability to avoid the utilization of larger metal crystals have been introduced as CT agents. This approach includes encompassing of small nanoparticulates (<6 nm) within a bigger nanoparticulate stabilized by amphiphiles (e.g., diblock copolymers or phospholipid surfactant) or noncrystalline organically soluble radio-opaque metallic small molecule agents (organometallics or metal complexes) [15]. Use of generally recognized as safe (GRAS) quality components such as lipids, vegetable oil, etc., makes them highly biocompatible agents [16]. Avoiding inclusion of larger metallic crystalline particles guarantees the rapid clearance of the residual metals through natural biliary renal excretion mechanisms. In addition to this, lipids and other innumerable chemical modifications used as outer coating in such soft nanoparticles offer high biocompatibility, homing, and drug delivery properties according to the nature of modifying agent [1].

5.3.1 Nanoparticles Incorporating Bismuth

Nanoparticles based on polyvinyl pyrolidone-coated Bi_2S_3 with a size of ~50 nm have been reported [17,18]. A significant enhancement in delineation and signal Hounsfield Unit (HU) = 560 was seen in the cardiac ventricles and major arterial and venous structures in rats. Although their circulation time (>2 h) was satisfactory for the purpose and in vivo sensitivity is adequate, their clinical translatability is questionable due to long-term safety concerns because of forbidden in vivo clearance of larger particulates. The long-term safety studies and toxicity evaluation will not be pursued here because of the presentation's preclinical nature (Figure 5.1).

Pan et al. engineered and evolved radio-opaque colloidal nanoparticles packed with small molecule organic–metal complexes at high density in lieu of large metal crystals [16]. Synthesis of bismuth-incorporated nano-K led to required high metal content (>500,000/NP) with a low molecular weight organo-soluble bismuth complex (Figure 5.1). These nano-Ks have a hydrodynamic diameter between 180 and 250 nm with a negative electrophoretic potential ranging from –20 to –27 mV. Outer membranes of these nano-Ks consisted of phospholipids and had efficiency of easily being functionalized for a variety of intravascular targets. It could be performed for ligand-directed homing to fibrin of microthrombus in coronary ruptured plaque. The special notice on particle size (~200 nm) throws light on its selectivity of homing. This property of the agent prevents its interaction with off-target intramural fibrin from within healed atherosclerotic plaque intramural hemorrhages and trammels the agent to the vasculature during early circulation and imaging. These nano-K particles are composed of bismuth neodecanoate at 40%–70% element v/v inside a sorbitan sesquioleate core matrix. These particles were noted to be appropriate for K-edge coronary imaging at the site of high distinguishability required with strong attenuation of plaque calcium from fibrin-targeted contrast signal (Figure 5.2).

A classic avidin–biotin coupling technique was used to prepare biotinylated nano-K or the control nanocolloid (i.e., no bismuth, ConNC) flagging well-characterized fibrin-specific monoclonal antibody (NIB5F3), which was used to

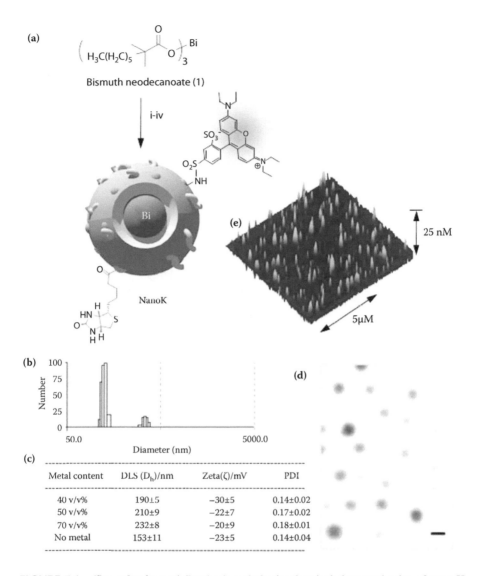

FIGURE 5.1 (See color insert.) Synthesis and physicochemical characterization of nano-K. (a) Schematic describing the preparation of bismuth-enriched K-edge nanocolloid (nano-K): (i) suspension of bismuth neodecanoate (1) in sorbitan sesquioleate, vigorously vortex and mix, filter using cotton bed, vortex; (ii) preparation of phospholipid thin film; (iii) resuspension of the thin film in water (0.2 m*M*); (iv) microfluidization at 4°C, 12,000 psi, 4 min, dialysis (cellulosic membrane, MWCO 20K). (b) Hydrodynamic particle size distribution from DLS; (c) characterization table for three replicates of nano-K; (d) anhydrous state transmission electron microscopy (TEM) images (staining: uranyl acetate; scale bar: 100 nm; (e) atomic force microscope (AFM) image.

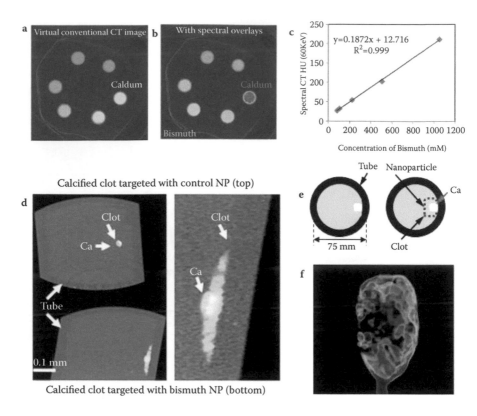

FIGURE 5.2 **(See color insert.)** Spectral CT cross-sectional slices (top) and gradient rendered images (below) of fibrin clots targeted with control (a, e) and nano-K replicates (b–d); integral bismuth distribution in axial slices of fibrin clots: Bound on bismuth layer thickness calculated with scanner spatial resolution at 100 mm, voxel size in reconstructed image (100 mm)³. Bismuth layer thickness: 1 or 2 voxels; bismuth surface density was calculated from integrations perpendicular to the surface layer corresponding to an average 3.5 mass% bismuth for a 100 mm layer thickness.

pretarget cellular fibrin clots in vitro [16]. The targeted fibrin clot samples were seen as K-edge images in cross section (Figure 5.2a), which generate three-dimensional maximum intensity projection reconstructions (Figure 5.2b). It showed excellent signal enhancement from the bismuth-enriched fibrin clot surface (bottom, far left), compare to the control clot (top, far left). Exposure to targeted nonmetallic nanoparticles had negligible contrast, revealing only the highly attenuating calcium. The specified size of nano-K, similar to prior examples in this phantom model, prevents the deep penetration into the tight weave of fibrin fibrils. The morphological description reveals the nano-K layer bound on the surface of the clots as 1 to 2 voxels (100 μm × 100 μm × 100 μm) thickness with an average density of 3.5 mass% of bismuth for a layer thickness of 100 μm.

The possibility of detecting fibrin (presented by unstable human atherosclerotic vascular tissue) at its physiological density was tested in carotid artery endarterectomy

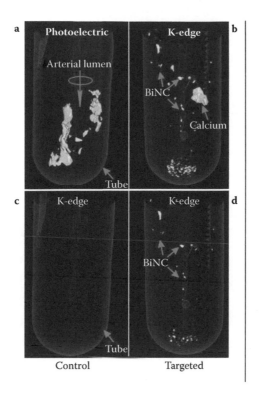

FIGURE 5.3 **(See color insert.)** (a) Photoelectric image of CEA specimen targeted with CoNC showing calcium (presented in red); (b) spectral CT image of CEA specimen targeted with CoNC revealing no presence of the K-edge metal; (c) photoelectric image of CEA specimen targeted with nano-K showing but not differentiating the attenuation contrast of both plaque calcium and the fibrin-targeted nano-K; (d) spectral CT image of CEA specimen after fibrin-targeted nano-K showing the spatial distribution and sparse concentration of thrombus remaining on human carotid specimen. Note: Fibrin-targeted nano-K signal at the bottom of the tube represents thrombus dislodged from CEA during processing.

(CEA) specimens. Fibrin-specific nano-K was used to expose microscopic fibrin deposits in CEA in vitro. Classic CT images (combination of Compton, photoeffect, and bismuth) of CEA tissue (ConNC or exposed to fibrin-targeted nano-K) revealed heavy calcification of both specimens but no differentiation of the nano-K particles from the calcium based solely on x-ray attenuation. On the other hand, bound fibrin-targeted nano-K (color: gold) at 90.5 keV distinguished the signal from the complicating calcium (color: red) at 4.0 keV using K-edge imaging (Figure 5.3).

A rabbit thrombus model was used to test the in vivo sensitivity of nano-K. Within the iliac artery of an atherosclerotic rabbit, a balloon overstretched injury-induced thrombus was formed. Nano-K particles directly decorated with antifibrin monoclonal antibody were incubated in situ with the thrombus for more than 30 min. The incubated system was kept to wash out the unbound material by keeping it in open circulation for 1 hour. The reconstructed intra-arterial thrombus enhanced with fibrin-bound nano-K was seen in spectral CT (Figure 5.4b). It was observed that the

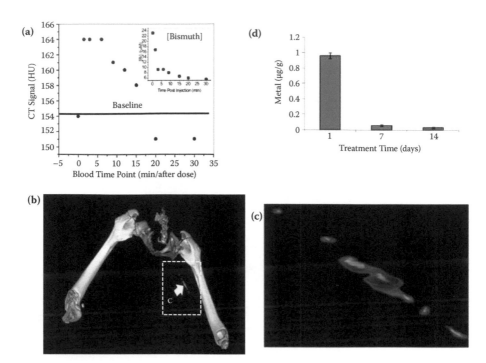

FIGURE 5.4 (a) CT blood pool signal in rabbits following IV injection of nano-K. CT scan imaging parameters were the following: thickness 0.8, increment 0.8, kv 90, mAs 1500, resolution HIGH, collimation 4 × 0.75, pitch 0.35, rotation time 1.5 seconds, FOV 75 mm. Inset: the concentration of bismuth (ICP) in blood versus time postinjection. Note that the background signal is at baseline in less than 30 minutes; (b, c) targeting in situ clot (thrombus) in rabbits; (d) 2-week clearance profile of bismuth from mice.

partial-occlusive thrombus was formed within a 1.41-mm-diameter artery comparable to a small coronary artery in humans. Signals from the pathology within the vessel from the attenuation effects of bone were clearly distinguishable from the bismuth signal obtained from nano-K in spectral CT. The traditional CT images were acquired simultaneously with K-edge image (Figure 5.4), indicating the huge potential of the technique in a clinical setting.

As essential criteria for a good CT agent, the heavy metals used for imaging must be cleared from the subject's body in a reasonably short time period. Inductively coupled plasma-mass spectrometry (ICP-MS) methodology was used to monitor and analyze the whole-body bioelimination of nano-K bismuth following intravenous injection. A 2-week study revealed that >98% metal was cleared within 7 days from the mice (Figure 5.4d). This is expressing the significant illustration of the concept of bioeliminating the heavy metal payload of nano-K within a reasonably short timeframe. Residual biodistribution of nano-K was examined on the 14th day into liver, spleen, and kidney, known as primary clearance organs. Analytical reports with ICP revealed <10 ppb mL^{-1} below the statistically valid lower detection limit of the instrument and method.

5.3.2 Nanoparticles Incorporating Gold

Gold (Au, 79) absorbs x-rays strongly to generate secondary electrons and photons that can kill neighboring cells (e.g., cancer cells). Gold can transform absorbed light into heat, resulting in localized temperature rises, which have recently been used to provide contrast for photoacoustic imaging or for photothermal therapy [19,20]. Gold with its high atomic number and higher absorption coefficient in opposition to iodine (I, 53) provides about 2.7 times greater x-ray contrast per unit weight. At 100 keV, gold produces 5.16 cm^2 g^{-1} in comparison to iodine, which is 1.94 cm^2 g^{-1}. This is particularly prominent at the higher x-ray tube voltages (i.e., 90, 120, and 140 kV) used in biomedical imaging. Furthermore, gold nanoparticles, offer good overall safety profiles and ease of functionalization to target biological markers. However, this advantage can be upturned for particles below 2 nm, which have unique chemical reactivity, or annulled through surface coatings and charge [19]. On the other hand, large particles of gold (>10 nm) are found to be safe in cell cytotoxicity studies but their lack of renal clearance in humans poses long-term safety questions. The crystalline structured form of large gold nanoparticles may compromise the validity of comparisons with other forms of medical gold, such as gold sodium thiomalate for rheumatoid arthritis, which has long fallen into unpopularity due to countless side effects.

Synthetic protocols for making gold nanoparticles are dominated by the chemical reduction methods in aqueous and nonaqueous media. A pioneering method developed by Turkevich is a gold chloride reduced with sodium citrate in aqueous solution at boiling temperature to produce a citrate-capped particle [21]. The particle sizes are tunable from 15 to 150 nm by simple adjustment of the ratio of the reducing agent. The stability of the particles in saline solutions can be driven by coating with organic or inorganic molecules (e.g., polyethylene glycol [PEG], silica, lipids, bovine serum albumin [BSA], and various polymers). Different coating methods have been adopted (e.g., ligand substitution, wrapping with amphiphiles, or embedding in a carrier matrix). In the Brust–Schiffrin method [22], a nonaqueous procedure is followed to produce organically coated gold nanoparticles.

A synthetic protocol for making gold nanocages following a galvanic replacement reaction was developed by Xia et al. [23]. Ag solid was used as a template for the reduction of $HAuCl_4$ due to the electrochemical potential difference to form Au atoms. In this reaction, the growth of AuNCs was templated over silver nanocubes that were concomitantly produced ($3Ag$ (s) + $AuCl_4^-$(aq) → Au (s) + $3Ag^+$ (aq) + $4Cl^-$ (aq)). Ag nanocubes were produced using $AgNO_3$ or CF_3COOAg as a precursor to Ag and a polyol reduction by using ethylene glycol as a solvent and a source of reducing agent. This reaction was further assisted by using a capping agent, poly(vinyl pyrrolidone) (PVP). Subsequently, the addition of catalytic amounts of NaHS and or Cl^- ions imparted rapid nucleation of single-crystal seeds while, concurrently, due to oxidative etching, discarded twinned seeds form single-crystal Ag nanocubes. Additional titration of these Ag nanocubes with $HAuCl_4$ followed at 100°C to ensure the maximal growth of thin-layered Au on the surface of the Ag crystals and to avoid precipitation of AgCl. Solvent-less "green" approaches such as microwave and UV irradiation were also followed to prepare these particles [24,25].

Several of these gold nanoparticles were studied in biological systems. Prostate-specific membrane antigen (PSMA) specific aptamers have been conjugated to gold nanoparticles for targeting prostate cancer cells in vitro. Kim et al. [26] demonstrated the detection of AuNPs with a clinical CT scanner. Chemotherapeutic doxorubicin was coupled to these particles to study the therapeutic benefits of these particles. A dual modality probe from Au nanoparticles coated with a Gd chelate was developed [27]. Comparative analyses of both CT and MR images of a rat with mixed Au-Gd nanoparticles showed augmentation in the kidneys, bladder, and urine. CT and MRI results were corroborated ex vivo with inductively coupled plasmon resonance mass spectrometric analyses of the excised tissues. Gold nanorods (AuNR) were synthesized by the Kopelman group and attached to the UM-A9 antibodies for specifically targeting squamous cell carcinoma (SCC) head and neck cancer [28]. In this proof-of-principle experiment, the feasibility to identify the existence of SCC cancer cells through CT scans was demonstrated. Lisinopril was used as a targeting moiety to prepare gold nanoparticle-based functional CT contrast agents following a ligand exchange reaction on citrate-coated gold nanoparticles [29]. Targeted gold nanoparticles assessed the targeting of angiotensin converting enzyme (ACE) using conventional CT. Clear enhancement was seen in the region of the lungs and heart, indicating the successful accumulation of gold particles to ACE. Cormode et al. developed a gold high-density lipoprotein nanoparticle contrast agent (Au-HDL) for characterization of macrophage burden, calcification, and stenosis of atherosclerotic plaques [30]. A spectral CT technique was employed to image these nanoparticles nonspecifically accumulated in macrophages in atherosclerotic plaques. For an atherosclerosis animal model, apolipoprotein E knockout (apoE-KO) mice were used. Gold nanoparticles were administered intravenously (500 mg/kg). The blood pool iodinated contrast was injected 24 hours later followed by the spectral CT imaging. Specific accumulation of the nanoparticles in macrophages was further established by transmission electron microscopy and confocal microscopy and histopathology.

Gold nanobeacons (GNBs) are another attractive choice developed for conventional and spectral CT imaging. The synthesis of GNBs followed a "particle within particle" approach [31]. Small (3–4 nm sized) octanethiol-coated gold nanoparticles were encapsulated (nominally 6,000 gold atoms/nanobeacons) within a bigger, vascularly constrained nanoparticle (~120 nm) stabilized by phospholipids. The GNB_1 particles were 154 ± 10 nm with polydispersity and zeta potential of 0.08 ± 0.03 and -47 ± 7 mV, respectively. ICP-MS determined the gold content, which was 1080 μg/g of the 20% colloid suspension. These nanoparticles showed excellent properties for sensitive photoacoustic tomography (PAT) imaging of atherosclerosis, sentinel lymph nodes, and, for the first time, detecting nascent sprouting angiogenic blood vessels. However, preliminary results with GNB particles produced poor contrast enhancement even in phantoms by computed tomography. The result was expected that, as for effective molecular imaging, nanoparticles designed for CT imaging must incorporate very high concentrations of elements. Phantom experiments suggested that at least 5×10^5 metal atoms per nanoparticle (e.g., gold atoms, 200 nm agent) are required for robust in vivo CT imaging.

The issue was later solved by Schirra et al. by the successful development of a second-generation gold nanobeacon (GNB$_2$). GNB$_2$ integrated at least five times more metal (nominally ~600,000 gold atoms/beacons) than the previous generation [32]. The particles were produced in a similar manner as lipid-encapsulated, vascularly constrained (>160 nm) nanobeacons with tiny, organically soluble oleate-coated gold nanoparticles (3–4 nm). Higher gold encapsulation was achieved by replacing the core material from vegetable oil to polysorbate (Figure 5.5). Polysorbate acts as a secondary surfactant, which helps to stabilize the higher metal loading. Anhydrous state transmission electron microscopy images revealed the presence of tiny gold particles sheathed within an outer lipid monolayer (Figure 5.6).

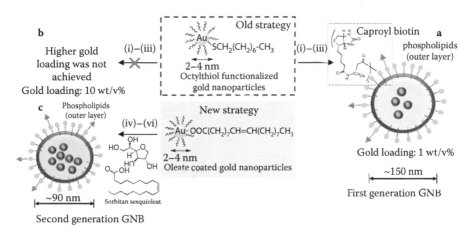

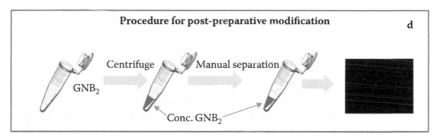

FIGURE 5.5 (**See color insert.**) Synthesis and postpreparative modification of GNB. (a) Previous synthetic approach to first-generation GNB particles from octanethiol-coated AuNP with a final gold loading ca. 2 w/v%: (i) gold nanoparticles suspended in vegetable oil; (ii) preparation of phospholipids thin film; (iii) microfluidization of gold nanoparticle–vegetable oil with surfactants in water. (b) New strategy to increase final gold loading up to 10 w/v%: Oleate-coated gold nanoparticles were suspended in sorbitan sesquioleate and microfluidized with phospholipids thin film mixture as a 2-0-20 formulation. Reaction condition: (i) gold nanoparticles suspended in polysorbate; (ii) preparation of phospholipids thin film; (iii) microfluidization of gold nanoparticle–polysorbate with surfactants in water; 20,000 psi (141 MPa), 4°C, 4 min; (d) schematic representation of the procedure to concentrate GNB2 prior to spectral CT imaging.

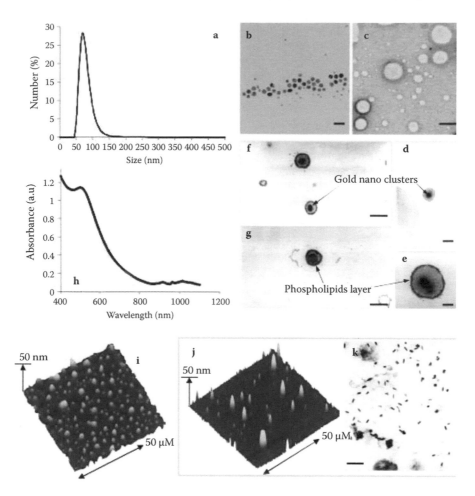

FIGURE 5.6 **(See color insert.)** Characterization of GNB. (a) Number-averaged hydrodynamic diameter of GNB2 from dynamic light scattering measurements; TEM images of (b) oleate-coated gold nanoparticles, scale bar = 20 nm; (c) control nanobeacons with no gold incorporated, scale bar = 100 nm; (d) first-generation GNB (scale bar = 100 nm); (e–g) second-generation GNB revealing the gold clusters entrapped inside the phospholipids membrane (scale bar = 100 nm); (h) UV-vis spectrum of GNB2 in water confirming the presence of gold nanoparticle cluster and their surface plasmon resonance. AFM images of GNB2 (i) and GNB (j, rod); (k) TEM image of GNB (rod) showing the presence of discrete gold rod nanoparticles of larger dimension within a phospholipid encapsulation (scale bar = 100 nm).

The use of statistical image reconstruction showed that high signal-to-noise ratio could allow dose reduction and/or faster scan times. In an in vivo model for sentinel lymph node (SLN) detection, second-generation gold nanobeacons injected into the foot-paw of a mouse migrated to a draining flank node where the accumulation produced a very strong spectral CT signal resolved from skeletal CT contrast interference (Figure 5.7).

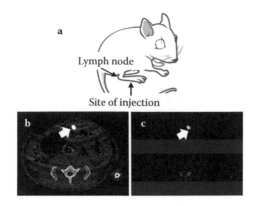

FIGURE 5.7   In vivo noninvasive spectral CT imaging of sentinel lymph nodes. (a) A cartoon illustrating the site of injection and the area of interest; 150 mL of nanobeacons were injected intradermally in all the cases. (b) Regional sentinel lymph nodes were clearly contrasted in conventional CT. (c) K-edge contrast of accumulated gold in the lymph node was selectively imaged with spectral CT.

5.3.3   NANOPARTICLES INCORPORATING YTTERBIUM

In the 1980s Unger et al. explored the use of ytterbium (Yb) as a CT contrast agent. It was demonstrated that Yb showed more radio opacity than equimolar concentration of iodine at 125 kVp [33]. An overall LD50 of ytterbium-DTPA chelate was reported to be 10 mM/kg (1.73 g ytterbium/kg) in rats. The construct was evaluated to be effective for pulmonary angiography in dogs. Krause et al. synthesized a more kinetically stable ytterbium chelate, Yb-EOB-DTP [34,35], which showed excellent tolerability of the complex in vitro (thromboplastin time, effect on erythrocytes) and in vivo (acute, neural, and cardiovascular toxicities) in rat and rabbit models. In a systematic comparative study of metal contrast enhancement, Zwicker et al. established a reduced enhancement in the order of Gd (120 kV) > Gd (137 kV) > Yb (120 kV) > Yb (137 kV) > iodine (120 kV) > iodine (137 kV) [36]. The specific signal augmentation from Gd was 40.8 (120 kV), 34.2 from Yb, and 29.6 HU of iodine. The in vivo dynamic CT imaging in dog showed augmentation of the liver in the order of 21 HU (Gd) to 19 HU (Yb) and 12 HU (iodine). Both Gd and Yb attained higher (190 vs. 157 HU) aortal contrast enhancement than iodine.

A simple bimodal imaging platform has been reported recently, based on PEG-coated NaYbF(4): Tm^{3+} nanoparticles for both CT and NIR-fluorescence bioimaging [37]. The nanoparticles were <20 nm in diameter and showed excellent in vitro and in vivo performances in the dual bioimaging with very low cytotoxicity and no detectable tissue damage. The high x-ray absorption coefficiency Yb^{3+} in the lattice of $NaYbF_4:Tm^{3+}$ nanoparticles helped it to function as a promising CT contrast medium while providing excellent sensitizing performance to enhance NIR-fluorescent emissions. The particles were excreted mainly via feces in rodents, without detectable remnant in vivo.

Liu et al. designed and synthesized a PEG-coated Yb_2O_3:Er nanoparticle suitable for both x-ray CT imaging and up-conversion imaging. When compared with an iodinated contrast agent (i.e., Iobitridol), these particles presented enhanced contrast at a clinical 120 kVp voltage. These PEG-UCNPs were facile to construct, possessed excellent stability against an in vivo environment, and held long blood circulation time. Cell-cytotoxicity assay, hemolytic activity, and postinjection histology analysis suggested good biocompatibility, indicating the feasibility of PEG-UCNPs for in vivo applications. By doping 5% Er(3+) into the nanoparticles, PEG-UCNPs presented a long-term stable and nearly single-band red up-conversion emission upon continuous irradiation with a 980 nm laser [38].

Pan et al. developed a novel spectral CT probe based on Yb [39]. The synthetic approach involved the use of organically soluble Yb(III) complex to produce nanocolloids of Yb of noncrystalline nature incorporating a high density of Yb (>500 K/ nanoparticle) into a stable metal particle (Figure 5.8). The resultant particles were constrained to vasculature (~200 nm) and were specific for binding fibrin in the ruptured atherosclerotic plaque. Nanoparticles exhibited excellent signal sensitivity, and the spectral CT technique uniquely differentiated the K-edge signal (60 keV) of Yb from calcium (bones) (Figure 5.9).

5.3.4 NANOPARTICLES INCORPORATING TANTALUM

Tantalum can also be used as a CT contrast agent. Bonitatibus et al. developed a water-soluble tantalum oxide-based CT contrast agent [40] derived from Ta_2O_5. These nanoparticles (<6 nm) were within the threshold range of renal clearance and were cleared from the blood within a few seconds following intravenous injection in rodents. In vivo study showed clear delineation of a rat's vena cava and abdominal aorta. A PEGylated version of similar particles was devised with a much improved blood circulation time of 3 h within the hydrodynamic size range (5–15 nm, before PEGylation) [41,42].

5.3.5 OTHER AGENTS

Kweon et al. reported the use of phospholipids bilayer and incorporated water-soluble iodinated compounds together with organo-soluble iodized oil to increase the iodine concentrations in the liposomes for robust in vivo CT imaging [43]. Despite these potentials, the likelihood to translate a liposomal CT contrast clinically is very meager since it involves careful preparation, complicated purification, and relative instability in biological media.

Iodinated contrast molecules can be entrapped covalently or noncovalently to polymer chains. However, the fundamental requirement to incorporate heavy payloads of iodine molecules makes it challenging to synthesize stable suspension of polymeric iodinated contrast agents. Particular attention must be paid to increase their shelf-life properties and prevent disintegration and clustering in vivo while preserving their robust detection with x-ray imaging. Pan et al. recently reported the synthesis and characterization of a colloidal iodinated polymer nanoparticle (~200 nm)

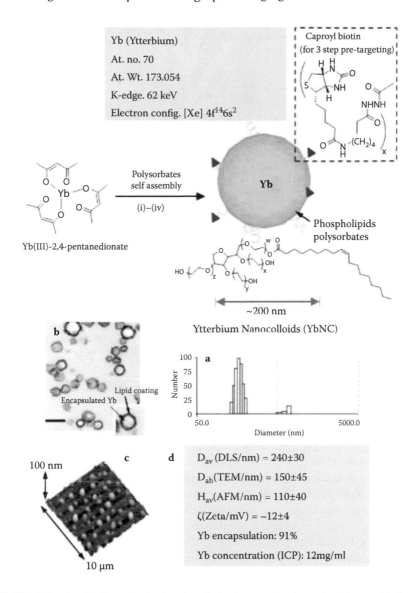

FIGURE 5.8 Synthesis and physicochemical characterization of self-assembled ytterbium nanocolloids (YbNC). Schematic describing the preparation of Yb-enriched YbNC: (i) Suspension of Yb(III)-2,4-pentanedionate in polyoxyethylene (20) sorbitan monooleate, vigorously vortex and mix, filter using cotton bed, vortex; (ii) preparation of phospholipids thin film composed of egg lecithin PC; (iii) resuspension of the thin film in water (0.2 m*M*); (iv) microfluidization at 4°C, 20,000 psi (141 MPa), 4 min, dialysis (cellulosic membrane, MWCO 20K); characterization table for a representative preparation of YbNC. (a) Number-averaged hydrodynamic diameter distribution of YbNC; (b) TEM images of the lipid-encapsulated nanocolloids; (c) AFM image of YbNC drop deposited over glass grid; (d) physicochemical characterization chart.

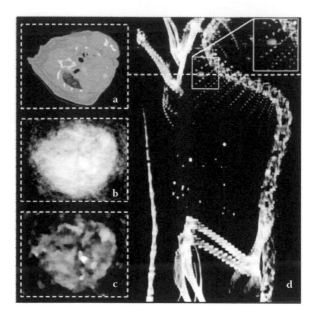

FIGURE 5.9 Blood pool imaging in mouse after bolus application of nontargeted Yb nano-colloids (6 mL/kg). (a) Pseudoconventional CT image composed from spectral measurements, slice through heart (dashed line). Statistical image reconstruction of Yb signal after 1 (b) and 20 iterations. (c) The volume-rendered conventional CT image with superpositioned Yb signal showing the accumulation of Yb in the heart (in box) and the clear separation from bone (d).

incorporating ethiodol stabilized by PS-b-PAA di block copolymer [15]. This report described a novel class of soft type, vascular-constrained, stable radio-opaque polymeric nanoparticle using organically soluble radio-opaque elements encapsulated by synthetic amphiphile. This agent offers several-fold CT signal enhancement in vitro and in vivo, demonstrating detection sensitivity reaching down to the low nanomolar particulate concentration range. A similar approach was followed by de Vries et al. to incorporate high iodine payload (130 mg I/mL) for blood pool CT imaging. This work demonstrated that polymer-stabilized particles remained stable similar to the lipid-stabilized emulsions in vivo [44].

5.4   CONCLUSION AND THE FUTURE OF CT IMAGING

CT is one of the most commonly pursued imaging techniques applied in clinics today. Improved signal sensitivity, rapid image acquisition, and faster reconstruction have made this technique even more powerful. Multidetector cardiac CTs (MDCT) can scan the heart within the span of a few beats, making it one of the most favored noninvasive approaches to assess coronary anatomy. However, MDCT has proven to be more useful for excluding coronary disease than for making positive diagnoses. The newest advancement in CT imaging technology, spectral (multicolored) CT uniquely enhances traditional CT images, which are based on the photoelectric and

Compton effects, with the capability to image and quantify certain metals based on distinctive K-edge values. Nanometer-sized agents are expected to play a critical part in the prospect of medical diagnostics owing to their capabilities of targeting specific biological markers, extended blood circulation time, and defined biological clearance. We discussed the fundamental design principles of nanoparticulate CT contrast agents with a special emphasis on the molecular imaging with noncrystalline high metal density nanobeacons. CT in combination with these molecularly targeted "soft" metal contrast agents can provide us quantitative information from the disease site, not only revealing the image of nanoparticle-enhanced pathology but also providing the amount of metal (i.e., the number of nanoparticles) bound in a given locality. In the cardiovascular area, in the future, quantification of the extent of intraluminal plaque rupture may allow risk stratification of lesions demanding a catheter-based stabilization protocol versus aggressive medical management of minor intimal ruptures expected to heal spontaneously. While spectral CT is still preclinical in nature, the advancement in photon-counting detectors and the development of dual energy CT instruments within hospitals across the globe will offer a preinstalled base to pursue clinical studies when these probes reach the clinical phase. The possibilities for molecular imaging with CT imaging are immense and may become crucial, but such realization will ultimately depend on the codevelopment of instrumentation and safer probes.

REFERENCES

1. Pan, D., Lanza, G. M., Wickline, S. A., Caruthers, S. D. 2009. Nanomedicine: Perspective and promises with ligand-directed molecular imaging. *European Journal of Radiology* 70 (2): 274–285.
2. Pan, D. 2013. Theranostic nanomedicine with functional nanoarchitecture. *Molecular Pharmacology* 10 (3): 781–782.
3. Jokerst, J. V., Gambhir S. S. 2011. Molecular imaging with theranostic nanoparticles. *Accounts of Chemical Research* 44 (10): 1050–1060.
4. Pysz, M. A., Gambhir, S. S., Willmann, J. K. 2010. Molecular imaging: Current status and emerging strategies. *Clinical Radiology* 65 (7): 500–516.
5. Kircher, M. F., Willmann, J. K. 2012. Molecular body imaging: MR imaging, CT, and US. Part I. principles. *Radiology* 263 (3): 633–643.
6. Shilo, M., Reuveni, T., Motiei, M., Popovtzer, R. 2012. Nanoparticles as computed tomography contrast agents: Current status and future perspectives. *Nanomedicine* (London) 7 (2): 257–269.
7. Köhler, T., Brendel, B., Roessl, E. 2011. Iterative reconstruction for differential phase contrast imaging using spherically symmetric basis functions. *Medical Physics* 38 (8):4542–4545.
8. Schirra, C., Roessl, E., Koehler, T., Brendel, B., Thran, A., Pan, D., Anastasio, M., Proksa, R. 2013. Statistical reconstruction of material decomposed data in spectral CT. *IEEE Transactions Medical Imaging* 32:1249–1257.
9. Feuerlein, S., Roessl, E., Proksa, R., Martens, G., Klass, O., Jeltsch, M., Rasche, V., et al. 2008. Multienergy photon-counting K-edge imaging: Potential for improved luminal depiction in vascular imaging. *Radiology* 249 (3):1010–1016.
10. Roessl, E., Herrmann, C., Kraft, E., Proksa, R. 2011. A comparative study of a dual-energy-like imaging technique based on counting-integrating readout. *Medical Physics* 38 (12): 6416–6428.

11. Roessl, E., Brendel, B., Engel, K. J., Schlomka, J. P., Thran, A., Proksa, R. 2011. Sensitivity of photon-counting based K-edge imaging in x-ray computed tomography. *IEEE Transactions in Medical Imaging* 30 (9): 1678–1690.

12. Liu, Y., Ai, K., Lu, L. 2012. Nanoparticulate x-ray computed tomography contrast agents: From design validation to in vivo applications. *Accounts of Chemical Research* 45 (10): 1817–1827.

13. Bloem, J. L., Wondergem, J. 1989. Gd-DTPA as a contrast agent in CT. *Radiology* 171 (2):578–579.

14. Zwicker, C., Hering, M., Langer, R. 1997. Computed tomography with iodine-free contrast media. *European Radiology* 7 (7): 1123–1126.

15. Pan, D., Williams, T. A., Senpan, A., Allen, J. S., Scott, M. J., Gaffney, P. J., Wickline, S. A., Lanza, G. M. 2009. Detecting vascular biosignatures with a colloidal, radio-opaque polymeric nanoparticle. *Journal of American Chemical Society* 131 (42): 15522–15527.

16. Pan, D., Roessl, E., Schlomka, J. P., Caruthers, S. D., Senpan, A., Scott, M. J., Allen, J. S., et al 2010. Computed tomography in color: Nano-K-enhanced spectral CT molecular imaging. *Angewandte Chemie* International Edition England 49 (50): 9635–9639.

17. Rabin, O., Manuel Perez, J., Grimm, J., Wojtkiewicz, G., Weissleder, R. 2006. An x-ray computed tomography imaging agent based on long-circulating bismuth sulphide nanoparticles. *Nature Materials* 5 (2): 118–122.

18. Leung, K. 2004–2013. Bismuth sulphide polyvinylpyrrolidone nanoparticles. Molecular Imaging and Contrast Agent Database (MICAD) [Internet]. Bethesda (MD): National Center for Biotechnology Information (US).

19. Pan, Y., Leifert, A., Ruau, D., Neuss, S., Bornemann, J., Schmid, G., Brandau et al. 2009. Gold nanoparticles of diameter 1.4 nm trigger necrosis by oxidative stress and mitochondrial damage. *Small* 5 (18): 2067–2076.

20. Pan, D., Pramanik, M., Wickline, S. A., Wang, L. V., Lanza, G. M. 2011. Recent advances in colloidal gold nanobeacons for molecular photoacoustic imaging. *Contrast Media Molecular Imaging* 6 (5): 378–388.

21. Kimling, J., Maier, M., Okenve, B., Kotaidis, V., Ballot, H., Plech, A. 2006. Turkevich method for gold nanoparticle synthesis revisited. *Journal of Physical Chemistry B* 110 (32):15700–15707.

22. Liz-Marzán, L. M. 2013. Gold nanoparticle research before and after the Brust–Schiffrin method. *Chemical Communications* (Cambridge). 49 (1): 16–18.

23. Xia, Y., Li, W., Cobley, C. M., Chen, J., Xia, X., Zhang, Q., Yang, M., et al. Gold nanocages: From synthesis to theranostic applications. *Accounts of Chemical Research* 44 (10): 914–924.

24. Dreaden, E. C., Alkilany, A. M., Huang, X., Murphy, C. J., El-Sayed, M. A. 2012. The golden age: Gold nanoparticles for biomedicine. *Chemical Society Review* 41 (7): 2740–2779.

25. Mieszawska, A. J., Mulder, W. J., Fayad, Z. A., Cormode, D. P. 2013. Multifunctional gold nanoparticles for diagnosis and therapy of disease. *Molecular Pharmacology* 10 (3): 831–847.

26. Kim, D., Jeong, Y. Y., Jon, S. 2010. A drug-loaded aptamer-gold nanoparticle bioconjugate for combined CT imaging and therapy of prostate cancer. *ACS Nano* 4 (7): 3689–3696.

27. Alric, C., Taleb, J., Le Duc, G., Mandon, C., Billotey, C., Le Meur-Herland, A., Brochard, T, et al. 2008. Gadolinium chelate coated gold nanoparticles as contrast agents for both x-ray computed tomography and magnetic resonance imaging. *Journal of American Chemical Society* 130 (18): 5908–5915.

28. Popovtzer, R., Agrawal, A., Kotov, N. A., Popovtzer, A., Balter, J., Carey, T. E., Kopelman, R. 2008. Targeted gold nanoparticles enable molecular CT imaging of cancer. *Nano Letters* 8 (12): 4593–4596.

29. Ghann, W. E., Aras, O., Fleiter, T., Daniel, M. C. 2012. Syntheses and characterization of lisinopril-coated gold nanoparticles as highly stable targeted CT contrast agents in cardiovascular diseases. *Langmuir* 28 (28): 10398–10408.

30. Cormode, D. P., Roessl, E., Thran, A., Skajaa, T., Gordon, R. E., Schlomka, J. P., Fuster, V., et al. 2010. Atherosclerotic plaque composition: Analysis with multicolor CT and targeted gold nanoparticles. *Radiology* 256 (3): 774–782.

31. Pan, D., Pramanik, M., Senpan, A., Allen, J. S., Zhang, H., Wickline, S. A., Wang, L. V., Lanza, G. M. 2011. Molecular photoacoustic imaging of angiogenesis with integrin-targeted gold nanobeacons. *FASEB Journal* 25 (3): 875–882.

32. Schirra, C. O., Senpan, A., Roessl, E., Thran, A., Stacy, A. J., Wu, L., Proska, R., Pan, D. J. 2012. *Materials Chemistry* 22 (43): 23071–23077.

33. Unger, E., Gutierrez, F. 1986. Ytterbium-DTPA. A potential intravascular contrast agent. *Investigative Radiology* 21 (10): 802–807.

34. Krause, W., Schuhmann-Giampieri, G., Bauer, M., Press, W. R., Muschick, P. 1996. Ytterbium- and dysprosium-EOB-DTPA. A new prototype of liver-specific contrast agents for computed tomography. *Investigative Radiology* 31 (8): 502–511.

35. Schmitz, S. A., Wagner, S., Schuhmann-Giampieri, G., Krause, W., Bollow, M., Wolf, K. J. 1997. Gd-EOB-DTPA and Yb-EOB-DTPA: Two prototypic contrast media for CT detection of liver lesions in dogs. *Radiology* 205 (2): 361–366.

36. Zwicker, C., Langer, M., Langer, R., Keske, U. 1991. Comparison of iodinated and non-iodinated contrast media in computed tomography. *Investigative Radiology* 26 (Suppl 1): S162–164.

37. Xing, H., Bu, W., Ren, Q., Zheng, X., Li, M., Zhang, S., Qu, H., et al. 2012. A NaYbF4: Tm3+ nanoprobe for CT and NIR-to-NIR fluorescent bimodal imaging. *Biomaterials* 33 (21): 5384–5393.

38. Liu, Y., Ai, K., Liu, J., Yuan, Q., He, Y., Lu, L. 2012. A high-performance ytterbium-based nanoparticulate contrast agent for in vivo x-ray computed tomography imaging. *Angewandte Chemie* International Edition England 51 (6): 1437–1442.

39. Pan, D., Schirra, C. O., Senpan, A., Schmieder, A. H., Stacy, A. J., Roessl, E., Thran, A., et al. 2012. An early investigation of ytterbium nanocolloids for selective and quantitative "multicolor" spectral CT imaging. *ACS Nano* 6 (4): 3364–3370.

40. Bonitatibus, P. J., Jr., Torres, A. S., Goddard, G. D., FitzGerald, P. F., Kulkarni, A. M. 2010. Synthesis, characterization, and computed tomography imaging of a tantalum oxide nanoparticle imaging agent. *Chemical Communications* (Cambridge) 46 (47): 8956–8958.

41. Bonitatibus, P. J., Jr., Torres, A. S., Kandapallil, B., Lee, B. D., Goddard, G. D., Colborn, R. E., Marino, M. E. 2012. Preclinical assessment of a zwitter-ionic tantalum oxide nanoparticle x-ray contrast agent. *ACS Nano* 6 (8): 6650–6658.

42. Torres, A. S., Bonitatibus, P. J., Jr., Colborn, R. E., Goddard, G. D., FitzGerald, P. F., Lee, B. D., Marino, M. E. 2012. Biological performance of a size-fractionated core-shell tantalum oxide nanoparticle x-ray contrast agent. *Investigative Radiology* 47 (10): 578–587.

43. Kweon, S., Lee, H. J., Hyung, W. J., Suh, J., Lim, J. S., Lim, S. J. 2010. Liposomes coloaded with iopamidol/lipiodol as a RES-targeted contrast agent for computed tomography imaging. *Pharmaceutical Research* 27 (7): 1408–1415.

44. de Vries, A., Custers, E., Lub, J., van den Bosch, S., Nicolay, K., Grüll, H. 2010. Block-copolymer-stabilized iodinated emulsions for use as CT contrast agents. *Biomaterials* 31 (25): 6537–6544.

6 Nanomedicine
Perspective and Promises

Dipanjan Pan

CONTENTS

This book is dedicated to the topic of molecular imaging and therapeutics, where we confer the role of functional nanoarchitecture. Over the past two decades, this multidisciplinary area of research has generated great vigor, showing high potential for clinical translational [1–4]. Innovation in chemistry, molecular biology, and engineering has shaped unique prospects for cross-disciplinary work in early detection and treatment of a disease at the molecular and cellular levels, enabling unparalleled safety and specificity [4–6]. The unprecedented potential of nanoparticles in imaging and drug delivery has been well proven [6–9]. Myriad progress has been made toward the development of defined nanostructure for performing multiple functions (e.g., imaging and therapy). Biological and biophysical obstacles are overcome by the incorporation of targeting ligands, contrast materials, and therapeutics into the nanoplatform, which allows for theranostic applications [10–12].

In this book, the readers have been introduced to the concept of theranostics, emphasizing the opportunities and challenges created by these soft materials for clinical translation. Size dictates the in vivo properties of these agents designed for targeting specific biological markers and strongly connects with its biodistributive nature, tissue accumulation, and cellular uptake [13–15]. There are already approved nanoparticles for medicinal applications and many are undergoing preclinical and clinical evaluation, including liposomes, polymeric micelles, dendrimers, Q-dots, gold-NPs TiO_2, etc. (Table 6.1) [13]. More than 200 companies are in the process of preparing nanotherapeutics. Most of these platforms are dominated by liposomal and polymer-drug conjugates, which account for greater than 80% of the total amount. Doxil is an example of a liposomal–PEG doxorubicin conjugate approved for clinical use. The encapsulation of the drug in PEGylated lipid vesicles prolonged drug circulation and improved drug accumulation in tumor tissue. In 1995, the FDA approved Doxil (Janssen Biotech) to treat an AIDS-associated cancer. Doxil is known to induce fewer side effects in comparison to its active chemotherapeutic ingredient, doxorubicin. However, it was later found that Doxil was unable to advance patients' survival rates compared with the parent doxorubicin. DepoCyt (liposomes-encapsulated cytarabine) is another liposomal-based drug carrier approved by the FDA for clinical use. The use of PEG as a polymeric drug

TABLE 6.1

List of Globally Approved Nanoparticles

Product	Nanoplatform/Agent	Indication	Status	Company
Doxil	PEGylated liposome/doxorubicin hydrochloride	Ovarian cancer	Approved 11/17/1995 (FDA50718)	Ortho Biotech (acquired by JNJ)
Myocet	Non-PEGylated liposomal doxorubicin nanomedicince	Metastatic breast cancer	Approved in Europe and Canada, in combination with cyclophosphamide	Sopherion Therapeutics, LLC, in North America and Cephalon, Inc. in Europe
DaunoXome	Lipid encapsulation of daunorubicin	First-line treatment for patients with advanced HIV-associated Kaposi's sarcoma	Approved in the United States	Galen Ltd.
ThermoDox	Heat-activated liposomal encapsulation of doxorubicin	Breast cancer, primary liver cancer	Received fast; track designation; approval expected by 2013	Celsion
Abraxane	Nanoparticulate albumin/paclitaxal	Various cancers	Approved 1/7/2005 (FDA21660)	Celsion
Rexin-G	Targeting protein tagged phospholipid/micro-RNA-122	Sarcoma, osteosarcoma, pancreatic cancer, and other solid tumors	Fully approved in Philippines; phase II/III (fast-track designation, orphan drug status acquired) in United States	Epeius Biotechnologies Corp.
Oncaspar	PEGylated asparaginase	Acute lymphoblastic leukemia	Approved 06/24/2006	Enzon Pharmaceuticals, Inc.
Resovist	Iron oxide nanoparticles coated with carboxydextran	Liver/spleen lesion imaging liver/spleen	In 2001, approved for the European market	Bayer Schering Pharma AG
Feridex	Iron oxide nanoparticles coated with dextran	Lesion imaging	Approved by US FDA in 1996	Berlex Laboratories
Endorem	Iron oxide nanoparticles coated with dextran	Liver/spleen lesion imaging	Approved in Europe	Guerbet

Reprinted with permission [13].

carrier was first tested in the early 1990s. These early studies clearly showed advantages of PEG to enhance the plasma stability, to increase the solubility of an insoluble drug, and to reduce immune response by "stealth" action. Abraxane is an early example of a polymer-bound drug (paclitaxel) conjugate (130 nm) approved by the FDA for the second-line treatment for breast cancer patients. The mechanism of action of Abraxane is believed to be EPR (enhanced permeability and retention)—in part through the transendothelial transport mechanisms via the albumin-binding protein gp-60 (Table 6.1).

Approximately 40 nanoformulations are being investigated at different levels of clinical development (Table 6.2). The majority of these agents rely on passive targeting approaches. A number of nontargeted nanoparticles are showing early promise in clinical trials and many others are being evaluated [16,17]. A few examples of these promising agents by National Cancer Institute (NCI) Alliance members are discussed next.

Investigators at Rice University developed nanoshells with a core of silica and a metallic (gold) outer layer (AuroLase). Once injected, nanoshells preferentially concentrate in cancer lesion sites due to their size. Nanoshells can further carry molecular conjugates to the antigens that are articulated on the cancer cells or in the tumor microenvironment. AuroLase therapy is being evaluated for the photothermal ablation of tumors showing early promise with greater efficacy of the therapeutic treatment and notably diminished side effects [18].

Clinical development of the [18F]-FAC family of PET imaging agents is being conducted by Sofie Biosciences. These agents are being tested for use in chemotherapeutics (e.g., gemcitabine, cytarabine, and fludarabine) and others to treat cancers including metastatic breast, nonsmall-cell lung, ovarian, and pancreatic, as well as leukemia and lymphomas. The technology was adopted from the laboratories of Drs. Caius Radu, Owen Witte, and Michael Phelps at the Nanosystems Biology Cancer Center (Caltech/UCLA CCNE). So far, there is participation from eight healthy volunteers in the biodistribution studies [19].

Dr. Mark Davis at the Caltech/UCLA CCNE translated two therapeutic technologies based on the cyclodextrin-based polymeric nanoparticle platform. Calando Pharmaceuticals is conducting clinical trials with this platform that encapsulate a small-interfering RNA (siRNA). An open-label, dose-escalating trial of their candidate agent is directed to understand the safety of this drug in patients resistant to other chemotherapies [20]. Another company (Cerulean Pharma, Inc.) is developing conventional chemotherapeutics (camptothecin/CPT) conjugated with the previously mentioned polymeric nanoparticles (CRLX101). An open-label, dose-escalation study of CRLX101 (previously named IT-101) is ongoing for solid tumor malignancies. This trial is also an open-label, dose-escalation study of CRLX101 (formerly named IT-101) administered in patients with solid tumor malignancies [21].

Drs. Gregory Lanza and Samuel A. Wickline at the Siteman Center of Cancer Nanotechnology Excellence (Washington University CCNE) initiated a clinical trial to study a nanoparticulate MRI contrast agent that binds to the $\alpha_v\beta_3$-intregrin found on the surface of the angiogenic blood vessels associated with early tumor development [22]. Ligand-directed perfluorocarbon nanoparticles were found to be effective acoustic contrast agents and subsequently helped to expand this platform technology

TABLE 6.2
Nanoparticle Cancer Therapeutics Undergoing Clinical Investigation

Product/Agent	Nanoplatform	Indication	Status	Company
Cyclosert	Cyclodextrin nanoparticles (cyclodextrin NP/SiRNA)	Solid tumors	Phase I	Insert Therapeutics (now Calando Pharmaceuticals)
CRLX101	Cyclodextrin NPs/caniptothecin PEGylated liposomal	Various cancers	Phase II	Cerulean Pharma
S-CKD602	CKD602 (topoisomerase inhibitor)	Various cancers	Phase I/II	Alza Corporation
CPX 1	Liposomal irinotecan	Colorectal cancer	Phase I	Celator Pharmaceuticals
CPX-351	Liposomal cytarabine and daunorubicin	Acute myeloid leukemia	Phase I	Celator Pharmaceuticals
LE-SN3S	Liposomal SN38	Colorectal cancer	Phase II	Neopharm
INGN-401	Liposo mal/FUS I polymeric nanoparticle	Lung cancer	Phase I	Introgen
NC-6004	(PEG-polyaspartate) formulation of cisplatin polymeric nanoparticle	Various cancers	Phase II	Nano Carrier Co.
NK-105	(PEG-polyaspartate) formulation of paclitaxel polymeric nanoparticle	Various cancers	Phase I	Nippon Kayaku Co. Ltd.
NK-911	(PEG-polyaspartate) formulation of doxorubicin	Various cancers	Phase I	Nippon Kayaku Co. Ltd.
NK-012	Polymeric micelle of SN-38	Various cancers	Phase II	Nippon Kayaku Co. Ltd.
SP1049C	Glycoprotein of doxorubicin	Various cancers	Phase II	Supratek Phanna Inc.
SPI-077	PEGylated liposomal cisplatin	Head/neck and lung cancer	Phase II	Alza Corporation

Name	Description	Indication	Phase	Company
ALN-VSP	Lipid nanoparticle formulation of siRNA	Liver cancer	Phase I	Alnylam Pharmaceuticals
OSI-7904L	Liposoma thymidylate synthase inhibitor	Various cancers	Phase II	OSI Pharmaceuticals
Combidex	Iron oxide	Tumor imaging	Phase III	Advanced Magnetics
Autimune	Colloidal gold/TNF	Solid tumors	Phase II	Cyt Immune Sciences
SGT-53	Liposome Tf antibody/p53 gene	Solid tumors	Phase I	SynerGene Therapeutics
BIND-014	PLGA/PLA NPs/Docetaxel	Prostate cancer and others	Phase I	BIND Biosciences
AuroLase	Gold-coated silica NPs	Head and neck cancer	Phase I	Nanospectra Biosciences
Rexin-G	Targeting protein tagged phospholipid micro-RNA-122	Sarcoma, osteosarcoma, pancreatic cancer, and other solid tumors	Phase II/III (fast track designation, orphan drug status acquired) in the United States; fully approved in Philippines	Epeius Biotechnologies Corp.
ThermoDox	Heat-activated liposomal encapsulation of doxorubicin	Breast cancer, primary liver cancer	Approved for breast cancer; phase III for primary liver cancer	Celsion
BIND-014	Polymeric nanoparticle formulation of docetaxel transferrin targeted	Various cancers	Phase I	BIND Bioscience
SGT53-01	Liposome with p53 gene P EG-glutaminase combined with glutamine	Solid tumors	Phase I	SynerGene Therapeutics
PEG-PCA and DON	Antimetabolite 6-diaio-5-oxo-l-norleucine (DON)	Various cancers	Phase I/II	EvaluatePharma
PEG-IFNfl2a	PEG-asys	Melanoma, chronic myeloid leukemia, and renal-cell carcinoma; melanoma, multiple	Phase I/II	Genentech
ADI-PEG20	PEG-arginine deiminase	Hepatocellular carcinoma	Phase I	Polaris

Reprinted with permission [13].

to include magnetic resonance tomography (SPECT) as well as therapeutic carriers in cancer, cardiovascular diseases, and rheumatoid arthritis [22]. The clinical trials with this particular agent were stalled due to the immune reaction caused presumably by the surface-present gadolinium–DOTA chelates.

At the MIT-Harvard Center for Cancer Nanotechnology Excellence, Dr. Ralph Weissleder is pursuing a study to determine if lymphotrophic superparamagnetic nanoparticles can be used to identify undetectable lymph node metastases [23].

Drs. Robert Langer and Omid Farokhzad of the MIT-Harvard CCNE developed targeted nanoparticles consisting of a polymer matrix, therapeutic payloads, surface-attached homing agents to facilitate accumulation in target tissue, avoidance of being cleared by immune system, and delivery of drug with desired release profile. BIND Biosciences is developing this technology. They initiated a phase 1 clinical study of its candidate agent (BIND-014) with an ascending, intravenous dose design to evaluate the safety, acceptability, and pharmacokinetics of patients with solid tumors. The primary objective of the study is to determine the maximum tolerated dose of BIND-014 and to assess preliminary evidence of antitumor activity [24].

Dr. Sanjiv Sam Gambhir (Stanford University CCNE) focused on the therapeutic response of carbon nanotubes (CNTs) to improve colorectal cancer imaging [25].

Although the EPR effect can enable nanoparticle transport in certain cancer tissues (e.g., inflammatory sites), most diseased tissues are not characterized by the remarkably leaky vasculatures. For these pathological circumstances, accumulation of nanoparticles will require an active mechanism of targeting. The reduction of uptake of nanoparticles by healthy tissues (also tissues rich with phagocytic cells) will require careful design that takes into consideration size, morphology, surface characteristics, etc. However, the selective recognition and delivery of nanoparticles in desired pathologic cells of interest will largely depend on active ligand-enabled homing. Selection of a homing agent for targeting is dependent on multiple critical variables including:

1. Identification of a receptor having required cell specificity, cell surface density, degree of internalization, and trafficking conduit
2. Identification of an agent with ample specificity for the biological receptor
3. Selective staging of the agent with or without a spacer to promote maximal projection of the ligand from the surface of the nanoparticulate

In most of the cases, developing "ideal" targeted nanoparticles for imaging and therapeutics will be contingent upon careful consideration of the physicochemical characteristics of the NP and the biology of the targeted tissue of interest. Refinement will be necessary to tailor the properties of these agents from initial proof of concept in in vitro, ex vivo, and in vivo studies.

As these platforms become more multifaceted with homing agents, drugs, stealth materials, and so on, it is critical now to reflect on the biocompatibility of components and the overall constructs. The National Characterization Laboratory (NCL), a federally funded US government facility, is geared to facilitate the biocompatibility study of nanomedicine platforms for clinical translation (Table 6.3). The most

TABLE 6.3

Nanomedicine Platforms Evaluated by NCL

Medicine	Indication	Particle Type	Company	Phase
PDS0101	Human papillomavirus-caused cancers	Positively chained liposome filled with antigen	PDS Biotechnology	Approved to begin phase I
Bind 014M	Prostate cancer	Tumor-targeting polymer nanoparticle filled with docetaxel	Bind Therapeutics	Approved to begin phase II
Cyt-6091	Solid tumors	Gold nanoparticle linked to tumor necrosis factor	Cytimmune Sciences	Phase II
AuroLase	Head and neck cancer, solid tumors	Gold nanoshells with silica core	Nanospectra Bioscience	Phase I
ATI-1123	Solid tumors	Liposome filled with docetaxel	Azaya Therapeutics	Phase I 1 complete
PNT2258	Non-Hodgkin's lymphoma and other cancers	Liposome filled with DNA interference fragment	Pronai Therapeutics	Phase II

Sources: Companies, NCL.
Note: NCL has done preclinical testing on six therapeutics now in clinical trials.

promising aspect is that major pharmaceutical companies, including AstraZeneca and Pfizer, are now slowly investing in nanotherapeutics evaluated by NCL [26, 27].

The past decade has seen an overabundance of nanotechnology-based approaches for theranostic application, biosensors, and real-time monitoring of circulating cancer and endothelial and bacterial cells. While nanotechnology offers great promise to address some of the burning issues in clinics today, the future of this technology in personalized medicine will largely be influenced by smart design principles for developing translatable, "safer" nanoplatforms and identifying novel receptors and high-affinity homing agents.

REFERENCES

1. Shin, S. J., Beech, J. R., Kelly, K. A. 2012. Targeted nanoparticles in imaging: Paving the way for personalized medicine in the battle against cancer. *Integrative Biology* (Cambridge) 5 (1): 29–42.
2. Pang, T. 2012. Theranostics, the 21st century bioeconomy and "one health." *Expert Reviews in Molecular Diagnosis* 12 (8): 807–809.
3. Lee, D. Y., Li, K. C. 2011. Molecular theranostics: A primer for the imaging professional. *American Journal of Roentgenology* 197 (2): 318–324.
4. James, M. L., Gambhir, S. S. 2012. A molecular imaging primer: Modalities, imaging agents, and applications. *Physiology Reviews* 92 (2): 897–965.
5. Pan, D., Lanza, G. M., Wickline, S. A., Caruthers, S. D. 2009. Nanomedicine: Perspective and promises with ligand-directed molecular imaging. *European Journal of Radiology* 70 (2): 274–285.

6. Alberti, C. 2012. From molecular imaging in preclinical/clinical oncology to theranostic applications in targeted tumor therapy. *European Review for Medical and Pharmacological Sciences* 16 (14): 1925–1933.

7. Wang, L. S., Chuang, M. C., Ho, J. A. 2012. Nanotheranostics—A review of recent publications. *International Journal of Nanomedicine* 7: 4679–4695.

8. Lammers, T., Aime, S., Hennink, W. E., Storm, G., Kiessling, F. 2011. Theranostic nanomedicine. *Accounts of Chemistry Research* 44 (10): 1029–1038.

9. Cabral, H., Nishiyama, N., Kataoka, K. 2011. Supramolecular nanodevices: From design validation to theranostic nanomedicine. *Accounts of Chemistry Research* 44 (10): 999–1008.

10. Kunjachan, S., Jayapaul, J., Mertens, M. E., Storm, G., Kiessling, F., Lammers, T. 2012. Theranostic systems and strategies for monitoring nanomedicine-mediated drug targeting. *Current Pharmacology and Biotechnology* 13 (4): 609–622.

11. Pan, D., Caruthers, S. D., Chen, J., Winter, P. M., SenPan, A., Schmieder, A. H., Wickline, S. A., Lanza, G. M. 2010. Nanomedicine strategies for molecular targets with MRI and optical imaging. *Future Medicinal Chemistry* 2 (3): 471–490.

12. Lanza, G. M. 2012. ICAM-1 and nanomedicine: Nature's doorway to the extravascular tissue realm. *Arteriosclerosis Thrombosis and Vascular Biology* 32 (5): 1070–1071.

13. Wang, R., Billone, P. S., Mullett, W. S. 2013. Nanomedicine in action: An overview of cancer nanomedicine on the market and in clinical trials. *Journal of Nanomaterials* Article ID 62968.

14. Albanese, A., Tang, P. S., Chan, W. C. 2012. The effect of nanoparticle size, shape, and surface chemistry on biological systems. *Annual Review of Biomedical Engineering* 14:1–16.

15. Elsabahy, M., Wooley, K. L. 2012. Design of polymeric nanoparticles for biomedical delivery applications. *Chemical Society Reviews* 41 (7): 2545–2561.

16. Zhang, L., Gu, F. X., Chan, J. M., Wang, A. Z., Langer, R. S., Farokhzad, O. C. 2008. Nanoparticles in medicine: Therapeutic applications and developments. *Clinical and Pharmacology Therapy* 83 (5): 761–769.

17. Wagner, V., Dullaart, A., Bock, A. K., Zweck, A. 2006. The emerging nanomedicine landscape. *Nature Biotechnology* 24 (10): 1211–1217.

18. Morton, J. G., Day, E. S., Halas, N. J., West, J. L. 2010. Nanoshells for photothermal cancer therapy. *Methods in Molecular Biology* 624: 101–117.

19. http://nano.cancer.gov/action/programs/caltech/

20. Davis, M. E., Zuckerman, J. E., Choi, C. H., Seligson, D., Tolcher, A., Alabi, C. A., Yen, Y., Heidel, J. D., Ribas, A. 2010. Evidence of RNAi in humans from systemically administered siRNA via targeted nanoparticles. *Nature* 464 (7291): 1067–1070.

21. Schluep, T., Hwang, J., Cheng, J., Heidel, J. D., Bartlett, D. W., Hollister, B., Davis, M. E. 2006. Preclinical efficacy of the camptothecin-polymer conjugate IT-101 in multiple cancer models. *Clinical Cancer Research* 12 (5): 1606–1614.

22. Pan, D., Lanza, G. M., Wickline, S. A., Caruthers, S. D. 2009. Nanomedicine: Perspective and promises with ligand-directed molecular imaging. *European Journal of Radiology* 70 (2): 274–285.

23. Shan, G., Weissleder, R., Hilderbrand, S. A. 2013. Upconverting organic dye doped core-shell nano-composites for dual-modality NIR imaging and photo-thermal therapy. *Theranostics* 3 (4): 267–274.

24. Hrkach, J., Von Hoff, D., Ali, M., Andrianova, E., Auer, J., Campbell, T., De Witt, D., et al. 2012. Preclinical development and clinical translation of a PSMA-targeted docetaxel nanoparticle with a differentiated pharmacological profile. *Science Translational Medicine* 4:128ra39.

25. Thakor, A. S., Gambhir, S. S. 2013. Nanooncology: The future of cancer diagnosis and therapy. *CA: Cancer Journal for Clinicians* doi: 10.3322/caac.21199. [Epub ahead of print]

26. http://cen.acs.org/articles/91/i35/Federal-Lab-Helps-Clients-Move.html

27. Crist, R. M., Grossman, J. H., Patri, A. K., Stern, S. T., Dobrovolskaia, M. A., Adiseshaiah, P. P., et al. 2013. Common pitfalls in nanotechnology: Lessons learned from NCI's Nanotechnology Characterization Laboratory. *Integrated Biology* (Cambridge) 5 (1): 66–73.

Index

T - #0438 - 071024 - C35 - 234/156/9 - PB - 9780367378509 - Gloss Lamination